STRATEGIES *for* SUCCESS

MATH Problem Solving

Strategies for Success: Math Problem Solving, Grade 5   OT113 / 324NA   ISBN-13: 978-1-60161-938-9
Cover Design: Bill Smith Group   **Cover Illustration:** Valeria Petrone/Morgan Gaynin

Table of Contents

Problem-Solving Toolkit

How to Solve Word Problems

Have you ever tried to get somewhere without clear directions? How did you find a way to get there? You can use the same kind of thinking when you are solving word problems.

Read the Problem Read carefully to be sure you understand the problem and what it is asking. Try to get a picture in your mind of what is going on and what is being asked.

Search for Information Look through all the words and all the numbers to see what information is given. Study any charts, graphs, and diagrams. Anything that might help you solve the problem is important.

Decide What to Do Think about the problem. If you are not sure how to solve it right away, ask yourself if you have solved a problem like this before. Think about all the problem-solving strategies you know. Choose one that you think will work.

Use Your Ideas Start to carry out your plan. Try your strategy. Think about what you are doing. Once in a while, ask yourself, *Am I on the right track?* If not, change what you are doing. There is always something else you can try.

Review Your Work Keep thinking about the problem. Finding an answer does not mean you are done. You need to keep going until you are sure you solved the problem correctly.

You can use the Problem-Solving Checklist on page 7 to make sure you have followed these important steps.

Problem-Solving Checklist

▢ Read the Problem

- ☐ Read the problem all the way through to get an idea of what is happening.
- ☐ Use context clues to help you understand unfamiliar words.

Ask Yourself

- ☐ How can I restate the problem in my own words?

▢ Search for Information

- ☐ Reread the problem carefully with a pencil in your hand. Circle the important numbers and math words.

Ask Yourself

- ☐ What do I already know?
- ☐ What do I need to find out to answer the question the problem asks?
- ☐ Does the problem have any facts or information that are not needed?
- ☐ Does the problem have any hidden information?
- ☐ Have I solved a problem like this before? If so, what did I do?

▢ Decide What to Do

- ☐ Choose a strategy that you think can help you solve the problem.
- ☐ Choose the operations you will use.

Ask Yourself

- ☐ How can I use the information I have to solve the problem?
- ☐ Will this problem take more than one step to solve?
- ☐ What steps will I use?

▢ Use Your Ideas

- ☐ Try the strategy you chose to solve the problem.
- ☐ Do the necessary steps.
- ☐ Write a complete statement of the answer.

Ask Yourself

- ☐ Do I need any tools such as a ruler or graph paper?
- ☐ Would an estimate of the answer help?
- ☐ Is my strategy working?

▢ Review Your Work

- ☐ Reread the problem.
- ☐ Check your computations, diagrams, and units.

Ask Yourself

- ☐ Is my answer reasonable? Does it make sense?
- ☐ Did I answer the questions the problem asks?

Problem-Solving Strategies

Look for a Pattern

You see ants crawling near the picnic table. There are hundreds of them!

But if you watch more closely, you can see a pattern. Some ants are going out to get food. Some ants are bringing food back to the ant colony.

Looking for patterns can help you understand nature. In a similar way, looking for patterns can help you solve math problems.

A stack of boxes at the store forms a pyramid. There is 1 box in the top layer, 4 boxes in the second layer, 9 boxes in the third layer, 16 boxes in the fourth layer, and so on.

Can you use a pattern to find the number of boxes in a stack 8 layers high?

Some patterns may not continue. Suppose you see 2 ant colonies on Monday, 4 on Tuesday, and 8 on Wednesday. Does that mean you will see 16 ant colonies on Thursday, and over a million three weeks from now?

Guess, Check, and Revise

Guessing and checking? Sounds like you start with an idea and you try it. If your idea does not work, then you try another one.

This strategy uses those same steps. But before you make your next guess, take some time to think. Revise your guess to get closer to the answer.

When you use the strategy "Guess, Check, and Revise," be sure to analyze the result of each guess to help you make your next guess.

The addresses on this side of Main Street are odd. Which two house numbers next to each other have a product of 143?

Start by trying a couple of simple numbers, such as 5 and 7. Find the product. $5 \times 7 = 35$. Your answer is much too low. So try 13 and 15. $13 \times 15 = 195$. This time your answer is too high.

As you continue making your guesses, you will learn how to get closer and closer to your answer.

Make a Table

It is easier to find places in a city when its streets are arranged in an organized way. Like organized streets, a table is useful, too. A table is organized in rows and columns, so you can find the information you need.

Tickets to a play cost $7 for a seat in the upper balcony, $12 for a seat in the lower balcony, and $25 for a seat on the main floor. There are 60 upper balcony seats, 40 lower balcony seats, and 200 main floor seats. How much money does the theater take in if a performance is sold out?

You can use a table to keep track of your calculations.

Seat Type	Cost of Each Seat	Number of Seats	Subtotal
Upper Balcony	$7	60	$60 \times \$7 = \quad \$420$
Lower Balcony	$12	40	$40 \times \$12 = \quad \$480$
Main Floor	$25	200	$200 \times \$25 = \$5,000$
		Total	$5,900

Do you see how the table makes it easier to find your way through the problem?

Use Logical Reasoning

A little bit of logical thinking can often save you a lot of time and work.

There are 128 teams in a tournament. If a team wins, it stays in the tournament. If it loses, it is out. How many tournament games must be scheduled?

You can draw or list all the match-ups, but both methods require a lot of work. Try to think logically. Eliminating 127 teams will take 127 games.

Each bracket is 1 game.

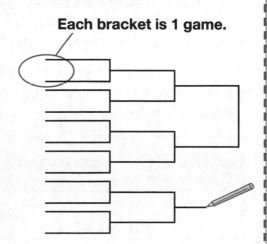

You may not need to do any computation at all to solve a math problem. A little logical thinking can go a long way.

Problem-Solving Strategies

Draw a Diagram

When you read a word problem, you may find yourself picturing in your mind what the problem is about. You can also draw a diagram to help picture what is happening.

The brick walkway around a community garden will be 3 feet wide. The garden is a rectangle 30 feet long and 20 feet wide. What will the area of the walkway be?

Do you see how drawing a diagram can help you picture the problem?

> You do not need to draw a complicated diagram.
> - Make sure the diagram shows the math in the problem accurately.
> - Use labels to help make the diagram clear.

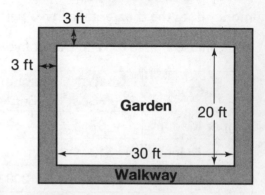

Solve a Simpler Problem

You and your neighbor are going to plant two kinds of vegetables in the community garden. You will choose from a list of 10 favorites. How many possible pairs of vegetables are there?

This problem would be easier if there were only 2 or 3 vegetables on the list. So start with simpler numbers to see if there is a pattern you can use to solve the problem.

Number of Vegetables to Choose From	Possible Pairs	Number of Pairs
2	AB	1
3	AB, AC, BC	3
4	AB, AC, AD, BC, BD, CD	6
5	AB, AC, AD, AE, BC, BD, BE, CD, CE, DE	10

> If a word problem seems complicated, think about how you can make it into a simpler problem. One way to make it simpler is to use simpler numbers. Simpler numbers can help you see how the original numbers are related.

Work Backward

You are on the third floor of the mall. You have to meet your parents at the entrance on the ground floor in 10 minutes. How will you get back to the entrance? You could go back the way you came, but to do this, you need to reverse the way you got to the third floor. For example, instead of going up the escalator, you will go down.

You can work backward to solve math problems, too. Try the number trick below.

> 1. Think of a number.
> 2. Double it.
> 3. Add 5.
> 4. Multiply the result by 3.
> 5. Subtract 15.
> 6. Divide by 6.
>
> Do you get the number you started with?

If you know the end result and how you got to it, you can work backward to find the beginning.

To find out how the number trick works, try working backward. Suppose you end up with 3. What number did you have before you divided by 6? What number did you have before you subtracted 15?

Do you see how to keep going backward until you get to the starting number?

Write an Equation

Take a look at the diagram of the garden and walkway. Does it look familiar? You need to find the area of the walkway.

First, you can find the area of one light section and the area of one dark section. Then you can write an equation to find the total area of the walkway.

Total area = 2 × area of light section + 2 × area of dark section

Can you take it from here?

An equation is like a set of directions. It is a short way to write mathematical steps. For example, think of "Area = length × width" as the set of directions to find the area of squares and rectangles.

Make a Graph

Sometimes, a picture can give you more information than a long description can. A graph is a picture of data. A graph can help you think about data in different ways.

```
                        X
              X    X    X
         X    X    X    X    X
    X    X    X    X    X    X    X
    +----+----+----+----+----+----+----+--->
    2    3    4    5    6    7    8
```
Possible Sums

In one game, two 4-sided playing pieces, each numbered 1 to 4, are tossed. The table shows the possible sums of the numbers tossed. The line plot shows the same data, but in a different way.

Numbers Tossed	Sum	Numbers Tossed	Sum
1, 1	2	3, 1	4
1, 2	3	3, 2	5
1, 3	4	3, 3	6
1, 4	5	3, 4	7
2, 1	3	4, 1	5
2, 2	4	4, 2	6
2, 3	5	4, 3	7
2, 4	6	4, 4	8

> A line plot can help you see how often something happens.

Make an Organized List

> Always check your list to make sure you did not leave anything out or list anything twice.

You buy a book and get $2.96 in change. What are all the different combinations of coins you could get as change? Finding all the different combinations may seem overwhelming. But you can make an organized list to help you.

Ms. Scott has students work in groups of three for many class projects. Each student has a role: Leader, Organizer, or Reporter. In how many different ways can these roles be assigned to three students? Make a list to find out.

Use L for Leader, O for Organizer, and R for Reporter.

Student	1 2 3	1 2 3	1 2 3
	R O L	O R L	L O R
	R L O	O L R	L R O

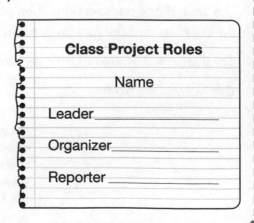

Class Project Roles

Name

Leader_____

Organizer_____

Reporter_____

Problem-Solving Skills

Check for Reasonable Answers

The thermometer reads 98°F, but it is snowing outside! Your common sense—and your cold fingers—tell you that something is not right.

Cold hands may not help you solve math problems, but your common sense can. It can let you know when you have made a mistake.

Suppose you need to find the change in temperature from 37°F to 63°F. To find out, you use your calculator to subtract 37 from 63. The calculator's display shows 100.

But the answer cannot be greater than 63. Do you see why?

> Whenever you solve a problem, look back to see if your answer is reasonable. Checking a problem to see if it is reasonable helps you spot mistakes and fix them.

Decide If an Estimate or Exact Answer Is Needed

To call your friend, you need her exact phone number, or else you will get the wrong person. With word problems, somtimes you need an exact answer to solve it. Sometimes, however, you only need an estimate.

> You know that an estimate can help you check an answer. But sometimes an estimate is all you need to answer the question.

Suppose you want to know which route from A to B is shorter. You do not need to find the exact distance of each route.

Do you see why one route must be more than 800 miles and the other must be less than 800 miles?

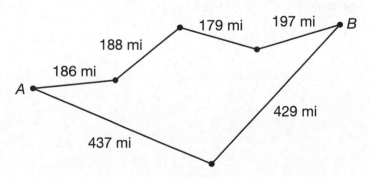

Find Hidden Information

Information you need is not always in plain sight. But if something is blocking your view, you must find a way to look around it.

Some math problems may also require you to use information that is not in the problem. The information may be in a diagram, in a table, or in a graph. It could be a fact not stated in the problem, such as *1 dollar = 100 cents.* The hidden information might even be a number you need to calculate.

Whatever form the hidden information takes, just think about what you need to do to find it.

> Read a problem carefully—and more than once—to be sure you have all the information you need.

Decide What Information Is Unnecessary for Solving

Have you ever noticed how many signs you see when you walk on a city street? You are not aware of them all because you ignore what is not important. You focus on what you are looking for. Focusing on what you need to find comes in handy for solving math problems.

West Street	East Street
	Park Entrance →

Suppose you want to know how to get to the park entrance. Which street do you need to take to get to the park?

> Solving a math problem can be like looking at street signs. You may not need to use all the information given.

Use Multiple Steps to Solve

You want to bike to your friend's house. But first, you need to get your bike. Next, you need to put your helmet on. Then you start pedaling. Most things need to be done step by step.

Sometimes, you need to do more than one step to solve a math problem, too.

Suppose you want to know if you can leave the park at 11 A.M. and get to the library by 11:15 A.M. You bike at about 15 miles per hour.

This problem has multiple steps. For the first step, you can use the map scale to figure out how far it is from the park to the library.

For the second step, you can use the distance you found and your speed to estimate how long it will take you to bike from the park to the library.

What will be your next step?

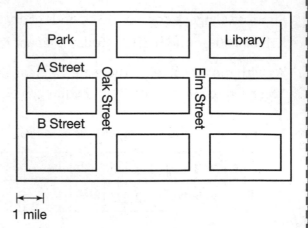

1 mile

To decide if you will need more than one step to solve a problem, first think about the data. Then think about the operations you may need to use to solve.

Interpret Answers

Take a look at this glass. Is it half empty, or is it half full? You interpret, or make sense of, what you see.

You interpret numbers in a word problem, too. After finding a number, you can ask yourself, *What does this number mean?*

Three friends want to share the 26 rocks they collected. How many rocks will each friend get?

You can divide: $26 \div 3 = 8\frac{2}{3}$

What does $8\frac{2}{3}$ mean in this situation? Can each person get $8\frac{2}{3}$ rocks? You need to interpret $8\frac{2}{3}$ so you can give an answer that makes sense.

You need to think about the problem situation when you interpret an answer.

Problem-Solving Skills

Choose Strategies

Suppose Ginger's dog eats $1\frac{1}{2}$ pounds of dry food each day. There is about half of a 40-pound bag left. Today is January 5. When will the food run out?

As with most problems, there are different ways to find an answer.
You need to choose the way, or strategy, you think will work best for you.

You might use the strategy *Make a Table*.

Date	Jan. 5	Jan. 6	Jan. 7	Jan. 8
Pounds Left	20	$18\frac{1}{2}$		

You might use the strategy *Write an Equation*.

Days left = Current amount of food ÷ Amount of food eaten each day

Days left = $20 \div 1\frac{1}{2}$

You might use the strategy *Draw a Diagram*.

Jan. 5   Jan. 6

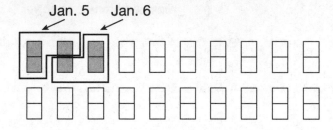

> A great thing about solving math problems is that you get to choose the way that works best for you.

Choose Operations

Many math problems can be simple and direct.
One operation may be all you need to solve them.

If a 40-pound bag of dry food costs $30, what is the cost for 1 pound?

Would you solve this problem by adding, subtracting, multiplying, or dividing? Which operation would you use?

> When you solve a problem, think before you compute. You may not even need to compute at all.

Solve Two-Question Problems

On many school buses, two children can ride in one seat.

Two questions are often in one word problem. When there are two questions in a problem, be sure to answer both of them.

A bus has 30 seats. Each seat can hold two students. Five seats are empty now, but the rest are filled. How many students can the bus hold? How many students are on the bus now?

Can you answer both questions?

When a word problem asks more than one question, take the time to think about each one and answer it before you move on to the next question.

Formulate Questions

Do you ever go through a day without asking a question? Probably not. People are naturally curious. So we ask a lot of questions.

Mathematicians often say that what matters in math is the kind of questions you ask. Your questions can be even more important than the answers.

As you look at the graph, you might wonder how many times faster a hawk moth flies than a honeybee.

What other questions might you ask?

Asking questions can be a good way to think about how to solve a problem.

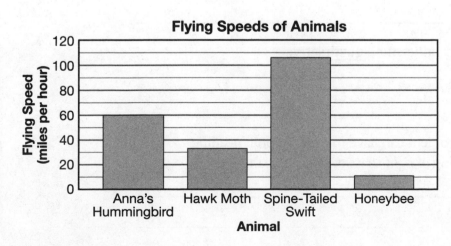

Flying Speeds of Animals

How to Read Word Problems

Word problems usually do three things.

- The words and numbers give you some information.
- The story helps you see how pieces of information are related.
- The problem asks a question or gives you a task.

Read to Understand

As you read, you may see words, facts, or symbols that you do not understand. Here are some things you can do to find out what they mean.

- You can **look up** the word, fact, or symbol.
- You can **use context clues** to help you understand.
- You can use **root words** to figure out the meaning.

Read the problem. Write the meaning of each underlined word. For items 2 and 3, explain your thinking.

> One leg of an <u>isosceles right triangle</u> is 4 feet long. What are the <u>dimensions</u> of the <u>quadrilateral</u> you can make with two of these triangles?

You need to read a word problem carefully to understand it. It often helps to read it again after you figure out what to look for.

1. I can look up <u>isosceles right triangle</u>.

   Meaning _____

   _____

2. I can use context clues to figure out the meaning of <u>dimensions</u>.

   Meaning _____

   _____

3. I can use a root word to understand <u>quadrilateral</u>.

   Meaning _____

   _____

Sometimes, a word in a problem can have more than one meaning. Compare the math meaning of *right* to its everyday meaning.

   Everyday Meaning _____

   Math Meaning _____

Look for Information

Read through a word problem once to be sure you understand what it is about.
Then read it again to identify numbers and words you need.

▶ A problem may have only the information you need.

> You are buying softballs for a school tournament.
> (Twelve softballs cost $39.99.) (Shipping costs $12.50.) How much
> will you pay for 36 softballs?

You do not need this information to solve the problem.

▶ A problem may have what you need and some extra details.

> You are buying softballs for a school tournament ~~with 16 teams~~.
> (Twelve softballs cost $39.99.) (Shipping costs $12.50.) How much
> will you pay for 36 softballs?

You need to know the meaning of "dozen."

▶ A problem may have only some of the information you need.
You must find a way to get the rest of the information.

> You are buying softballs for a school tournament with 16 teams. (One dozen)
> (softballs cost $39.99.) (Shipping costs $12.50.) How much will you pay for
> 36 softballs?

▶ Some information you need may be in tables, graphs, or diagrams.

**Read each problem. Study the information to decide how
it can help you solve the problem. Then write your answer.**

1. There are 552 students in Pat's school. The graph shows
   how many students take part in after-school activities.
   How many students participate in an after-school activity?

Information I can get only from the graph: _____

_____

Clubs: 105 | Music: 100 | None: 135 | Sports: 212

**Students in
After-School Activities**

2. The volume of this rectangular prism is 3,600 cubic centimeters.
   What is its length?

Information I can get only from the diagram: _____

_____

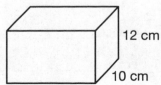

12 cm

10 cm

Mark the Text

Marking information you need in the text as you read can help you organize your thinking.

- You can circle numbers, including numerals and words.
- You can cross out information you do not need.
- You can underline the question the problem asks.
- You can mark something you need to look up or do not understand.

> I do not need this information.

Before Sam went shopping for clothes, ~~he saw two movies.~~ At one store, he bought (three T-shirts for $6.99 each), (two pairs of jeans for $24.99 each), and (a belt.) The total bill at that store was $77.96. How much did the belt cost?

> Here is the question I need to answer.

> This tells me that he paid for all the items at the same time.

Mark the text and tell why each mark is important.

The median age of students in Mika's class of 25 students was 10 years. The youngest student in the class was 8. When three new students joined the class, the median age changed to 11. What do you know about the ages of the new students?

I underlined _____

because _____

I crossed out _____

because _____

I circled _____

because _____

Decide What to Do

You can read a problem carefully to be sure you know what question you need to answer.

▶ Sometimes, a word problem clearly asks a question that you can answer by computing.

> You are buying lemonade for your two friends and for yourself. A glass of lemonade costs $2.25. <u>How much will you pay?</u>

Explain What operation will you use? Explain how you decided.

_____

_____

▶ Sometimes, you need to compute, but the result of your computation is not the answer to the problem.

> You are buying lemonade for your two friends and yourself. A glass of lemonade costs $2.25. You have $10.00. <u>Do you have enough money?</u>

Determine How can you find out if $10.00 is enough money?

_____

_____

▶ Sometimes, you do not need to compute at all. You can find the answer using a different method.

> In some states, there is sales tax only on items of clothing that cost more than $100 each. Mr. Lee bought five T-shirts for $8.99 each, two pairs of jeans for $25.99 each, and a winter coat for $119.99. <u>For which item did Mr. Lee have to pay sales tax?</u>

Apply You do not need to compute to find the answer. What do you need to do instead? What will your answer look like?

_____

_____

UNIT 1

Problem Solving Using Place Value and Whole Number Operations

Unit Theme:

Just for Fun!

What do you do for fun? In this unit, you will read about rides and attractions at an amusement park, puzzles and games people play, sporting events, and going to the movies. And of course, you will also find interesting ways to apply the math you already know.

Math to Know

In this unit, you will use these math skills:

- Use place value and number sense
- Add and subtract whole numbers
- Multiply and divide whole numbers

Problem-Solving Strategies

- Look for a Pattern
- Guess, Check, and Revise
- Make a Table
- Use Logical Reasoning

Link to the Theme

Finish the story. Include some of the facts from the table at the right.

Keri and her friend Ann are on their way to the Roaring Rapids Amusement Park. Ann has a booklet with facts about three new roller coasters at the park. She shows it to Keri.

New Roller Coasters

Name	Maximum Height (feet)	Top Speed (miles per hour)
The Escape	328	100
Titan	255	85
Top Thrill	400	120

_____

_____

_____

_____

_____

_____

_____

Use Math Language

Review Vocabulary

The list below shows vocabulary terms in this unit. Knowing the meaning of these terms will help you understand the problems.

difference   factor   pattern   quotient

divisor   multiple   product   remainder

Vocabulary Activity   Multiple-Meaning Words

Some terms have more than one meaning. Use terms from the list above to complete the following sentences.

1. I want to ride the roller coaster _____ times.

2. 12 is a _____ of 3.

3. It does not make any _____ to me which roller coaster I ride first.

4. The _____ between 400 feet and 255 feet is 145 feet.

Graphic Organizer   Word Map

Complete the graphic organizer.

- Write your own definition of *product*.

- Draw a picture or diagram to show what the term means.

- Write a number sentence that shows an example of the term.

- Write a number sentence that does not show an example of a product.

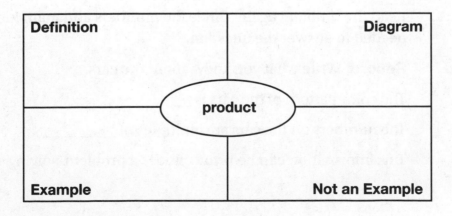

Strategy Focus
Look for a Pattern

MATH FOCUS: Place Value and Whole Number Operations

Learn About It

▪ Read the Problem

You can ride across Galaxy Fun Park in a space car. The space cars are shown below. Each car is numbered and painted. The numbers and colors follow a pattern. What color is car 350?

90	100	110	120	130	140	150	160	170	180
red	green	blue	red	green	blue	red	green	blue	red

Reread Ask yourself these questions as you read the problem again.

• What is the problem about?

_____

• What kind of information does the picture give me?

_____

• What am I asked to find?

_____

Mark the Text --→

▪ Search for Information

Read the problem again. Circle the numbers and math ideas needed to answer the question.

Record Write what you know about the cars.

The color pattern of the cars is _____ , _____ , _____ .

The numbers on the cars are multiples of _____ .

This information can help you choose a problem-solving strategy.

Decide What to Do

The cars have numbers and colors. You can see that both the numbers and the colors follow a pattern.

Ask How can I find the color of car 350?

- I can use the strategy *Look for a Pattern*.

- Since red is the first color in the pattern, I can start by finding the rule for the pattern of red cars.

- I can continue the pattern until I get close to the number 350. Then I can find the color of car 350.

Finding a pattern can help you do fewer calculations when you solve problems.

Use Your Ideas

Step 1 Find the number pattern of the red cars. The number on the next red car after car 180 is _____ .

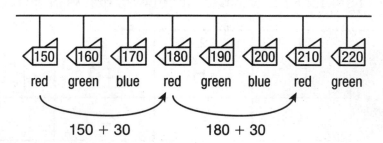

Write the pattern for the numbers of the red cars.
180, 210, 240, 270, _____ , _____ , _____

So the color of car 360 is _____ .

Step 2 Use the color of car 360 to find the color of car 350. What color comes before red in the pattern? _____

So the color of car 350 will be _____ .

Review Your Work

You know car 170 is blue. Count up by 30s from 170 to check if 350 is on your list. If it is, you know car 350 is blue.

Describe How did the diagram help you solve the problem?

_____

_____

Try It

Solve the problem.

(1) At an amusement park, a building like this one will be made of huge blocks. It will light up 4 times each minute. How many blocks are needed to make 8 layers?

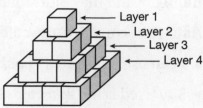
Layer 1
Layer 2
Layer 3
Layer 4

Mark the Text

▣ Read the Problem and Search for Information ·······

Mark the numbers and words that will help you.

▣ Decide What to Do and Use Your Ideas ···············

You can use the strategy *Look for a Pattern*.

Ask Yourself

Is the number of blocks from one layer to another increasing the same way?

Step 1 Look for a pattern in the number of blocks in each layer.

Layer	1 □	2 ⊞	3	4
Number of Blocks	1	4		

The number of blocks in each layer is the layer number times itself.

Step 2 Extend the pattern.

How many blocks are in the eighth layer? _____

Step 3 Add the number of blocks in each layer.

$1 + 4 + 9 + 16 + 25 + 36 +$ _____ $+$ _____ $=$ _____

So _____ blocks are needed.

▣ Review Your Work ·······························

Be sure to check your computations.

Identify What information is *not* needed to answer the question?

_____

_____

Apply Your Skills

Solve the problems.

(2) A theme park has a fireworks show every night. The show ends with a triangle of exploding stars, like the one below. How many stars will be needed for a triangle display made of 10 rows?

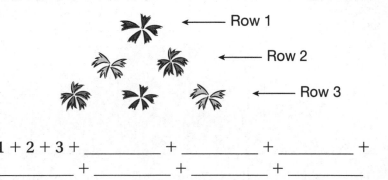

Row 1

Row 2

Row 3

1 + 2 + 3 + _____ + _____ + _____ + _____ + _____ + _____ + _____ + _____

Ask Yourself

How many stars will be in the fourth row?

Hint Draw the next few rows to see the pattern.

Answer _____

Recognize Marcos thinks that 10 stars will be needed. What did he forget to do?

_____

_____

(3) A new ride just opened at the amusement park. The first day, 75 people went on the ride. The ride was popular and word spread. On the second day, 20 more people went on the ride than the day before. If the number of people who go on the ride increases by the same number each day, on what day will the total number of riders reach 575?

On Day 2, _____ people went on the ride.

On Day 3, _____ people went on the ride.

Ask Yourself

What was the total number of riders for the first 2 days? The first 3 days?

Hint The problem asks about the total number of riders, not the riders on one day.

Answer _____

Determine How could you arrange the information to help you see a pattern?

_____

_____

④ All 91 characters at a theme park will march in a parade at noon. When they march, they will form a triangle in a pattern like the one shown. How many rows can they form?

Row Number	1	2	3	4	5
Number of Characters					

Answer _____

Discuss How did you use the strategy *Look for a Pattern* to solve the problem?

_____

_____

⑤ The Rocking Rocket ride swings right and left with longer and longer swings. It begins by swinging 20 feet to the right. Next, it goes 30 feet to the left, then 40 feet to the right, then 50 feet to the left, and so on. How far will the Rocket travel on its tenth swing if it continues this pattern?

Swing Number					
Swing Length (feet)					

Answer _____

Analyze What would be the direction of the tenth swing? How can you use even and odd numbers to determine the answer?

_____

_____

On Your Own

Solve the problems. Show your work.

6 A show at a theme park starts with a colony of 5 aliens. The number of aliens increases every 10 seconds, as shown in the diagram. The show lasts 2 minutes. How many aliens will be in the colony at the end of the show?

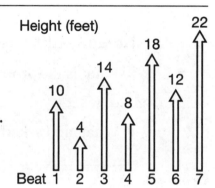

At start   After 10 seconds   After 20 seconds

Answer _____

Relate How did you know where to stop your pattern?

_____

_____

7 A water display shoots a colored wall of water into the air on every drum beat. On the first beat, the wall shoots 10 feet high. On the second, it shoots 4 feet high. Then it shoots 14 feet, then 8 feet, and so on. If it continues to change in this way, on which beat will the wall shoot 50 feet high?

Height (feet)

22
18
14
10   12
    8
4

Beat 1 2 3 4 5 6 7

Answer _____

Explain How can looking at every other beat make it easier to find the answer?

_____

_____

Create Look back at Problem 5. Change the swing number from *tenth* to some greater number. Write a new problem that can be solved using the strategy *Look for a Pattern*. Solve your problem.

Strategy Focus
Guess, Check, and Revise

MATH FOCUS: Addition and Subtraction

Learn About It

■ Read the Problem ·

In a game called Tossy Turvy, you throw 15 balls and try to get the highest score. Ana says she got exactly 108 points in one game. How many bull's-eyes and how many gutter balls did she get?

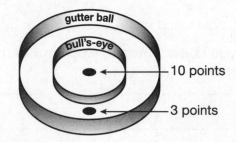

gutter ball
bull's-eye
10 points
3 points

Reread Ask yourself these questions as you reread.

- What is the problem about?

_____

- What am I asked to find?

_____

Mark the Text

■ Search for Information ·

Read the problem again. Circle numbers you will use to solve the problem.

Record What information will help you solve the problem?

Ana threw _____ balls.

Ana scored _____ points.

A bull's-eye is worth _____ points.

A gutter ball is worth _____ points.

You need to find a strategy that will help you use these numbers to solve the problem.

Decide What to Do

You know Ana's total score. You know how many points different tosses are worth. You know the number of tosses made.

Ask How can I find a combination of throws with a total score of 108?

- I can use the strategy *Guess, Check, and Revise.*

- For the 2 toss types, I can guess any 2 numbers that add to 15. Then I can check to see if these numbers are the right answer.

Use Your Ideas

Step 1 Guess numbers with sums of 15. Check to see if any are the right answer. Record each guess.

Guess	Number of Each Toss Type	Points from Each Type	Total Points	Does Total = 108?
1	10 bull's-eyes 5 gutter balls	_____ points _____ points		No, _____ is too _____.
2	8 bull's-eyes _____ gutter balls	_____ _____		No, _____ is too _____.
3	_____ bull's-eyes _____ gutter balls	_____ _____		
4				

Step 2 Revise to raise the total. Try fewer gutter balls and more bull's-eyes.

Step 3 Keep going until you have the correct total.

Ana got _____ bull's-eyes and _____ gutter balls.

Review Your Work

Check that the number of tosses and points add up to the numbers in the problem.

Clarify How do wrong guesses help you save time in finding the answer?

_____

_____

> When the calculations involve simple numbers, *Guess, Check, and Revise* can be a good strategy.

Try It

Solve the problem.

(1) These are the rules for the game *Got Your Number*. Suppose you are playing with your friend. Your friend ends with the number 160. What number was your friend thinking of at the start?

Got Your Number

1. Think of a number.
2. Add 25.
3. Double the result.
4. Subtract 40.
5. Say the number you end with.

Mark the Text

■ Read the Problem and Search for Information

Identify what the problem asks you to find. Underline the words and numbers that show how the starting number is changed.

■ Decide What to Do and Use Your Ideas

You can use the strategy *Guess, Check, and Revise*.

Fill in the table as you work through the steps below.

Guess	Starting Number	Steps			Result
		Add 25	Double	Subtract 40	
1	100	125	250	210	too high
2	50				too _____
3					

Ask Yourself

How can I organize my guesses in a table?

Step 1   Guess a starting number and check if it works.

Step 2   Use that information to make a better guess. Keep going until you get a starting number that works.

My friend's starting number was _____ .

■ Review Your Work

Start with your answer. Do all the steps. See if you end with 160.

(Explain)   Why should your starting number for the third guess be greater than 50?

_____

_____

Apply Your Skills

Solve the problems.

(2) Three friends play a game in which they earn tickets that they trade for prizes. Hank earns 15 more tickets than Dean and 23 fewer tickets than Trina. Together they earn exactly enough for a 200-ticket prize. How many tickets does Hank earn?

Hank	Dean	Trina	Total	= 200?
50	35	73	158	too low
80				

◀ **Hint** A guess of 50 for Hank is too low. Try a greater number next.

Ask Yourself

How should I revise my guess?

Answer _____

Identify How did guesses of 50 and 80 help you decide on your next guess?

_____

_____

(3) Larry scored a total of 42 points playing a basketball game. He gets 6 points for making a long shot and 2 points for making a short shot. He loses 3 points if he misses the shot altogether. Larry shoots the basketball 12 times. He makes exactly 3 short shots. How many long shots did he make?

◀ **Hint** Points are added to Larry's score when he makes shots and subtracted when he misses them.

Ask Yourself

Which numbers in the table always stay the same?

Long Shots		Short Shots		Miss		Total Score
Made	**Points**	**Made**	**Points**	**Misses**	**Points**	
4	+24	3	+6	5	−15	15
6						

Answer _____

Examine How would solving this problem be different if you were *not* given the number of short shots Larry made?

_____

_____

④ In the game *StopGap*, you put together lengths of pipe in order to fill a gap in the pipeline. In this round, the gap is 95 feet long and you have the 6 pieces shown in the table. Which of the pieces can you use to fill the gap exactly?

Your 6 Pipe Pieces	
5 feet	25 feet
10 feet	43 feet
15 feet	75 feet

Ask Yourself

Should I start with the longest piece or the shortest piece?

Hint Try using different numbers of pieces in each guess.

Guess	Pipe Lengths (feet)	Total	Is Total 95?
1	75 + 10 + 5		
2			

Answer _____

Determine Which length of pipe *cannot* be used? Explain.

_____

_____

⑤ Faye has earned 399 points on a quiz show. She needs exactly 500 points to win. Easy questions are worth 12 points, medium questions are 25 points, and hard questions are 40 points. How can Faye win the game by answering 5 questions correctly?

Ask Yourself

How many more points does Faye need to win?

Hint Start by determining the value of 5 hard questions and 5 easy questions.

Guess	Easy	Medium	Hard	Total Points = 500?
1				
2				

Answer _____

Analyze How can you explain the importance of checking your work before making another guess?

_____

_____

On Your Own

Solve the problems. Show your work.

6 In a game of bean bag toss, Jake, Olga, and Hugo scored 360 combined points. Olga scored 5 points more than Jake. Hugo scored 5 points more than Olga. What did each person score?

Answer _____

Appraise One student made a first guess of 50 + 55 + 60. Do you think that is a good first guess? Explain.

_____

_____

7 Dina is playing a card game in which each player gets seven cards. Cards are worth either 3, 6, or 11 points. Dina needs exactly 51 points. How many of each card does she need?

Answer _____

Extend Is it possible for Dina to get a total of 75 points with seven cards? Why or why not?

_____

_____

Create Look back at Problem 4. Change the length of the gap and the lenghts of the pipe pieces that can be used. Write a problem that can be solved using the strategy *Guess, Check, and Revise*. Solve your problem.

Strategy Focus
Make a Table

MATH FOCUS: Multiplication

Learn About It

Read the Problem

> Cindy's basketball team was losing a championship game by two points. Her team had scored 23 two-point baskets and 5 one-point free throws. At the buzzer, Cindy scored a three-point basket to win the game. What was the team's final score?

Reread Read the problem again. Answer these questions as you read.

- What is the problem about?

_____

- What kind of information is given?

_____

- What am I asked to find?

_____

Mark the Text

Search for Information

Read the problem again. Cross out information you do not need. Circle the numbers you think you will need.

Record What details do you know that will help you decide how to solve the problem?

You know how many baskets of each type the team made.

_____ two-point baskets

_____ one-point free-throws

_____ three-point baskets

You can use this information to choose a problem-solving strategy.

Decide What to Do

The details tell you this is a problem with two kinds of data: *number of baskets made* and *points each type of basket is worth*.

Ask How can I find the team's final score?

A table can help you organize the details.

- I can use the strategy *Make a Table* to organize the two kinds of data.

- I need to find the points scored for each type of basket. Next, I can multiply. Then I will add to find the total points scored.

Use Your Ideas

Step 1 Make a table to organize the information.

Find the product of the number of baskets and the number of points for each type.

Record each result in the table.

Type of Basket	Number of Baskets	Number of Points for Each Type	Points Scored
two-point basket	23	2		← 23 × 2 = ____
one-point free throw	5	1		← 5 × 1 = ____
three-point basket	1	3		← 1 × 3 = ____
		Sum →		

Step 2 Add the total points scored for each type of basket to find the total points scored in the game.

So the team's final score was _____ points.

Review Your Work

Check that your answer is what was asked for in the question.

Explain Suppose a student found the answer to be 23 + 5 + 1 = 29. What error might have been made and how can it be fixed?

_____

_____

Try It

Solve the problem.

(1) The booster club for a basketball team buys 75 T-shirts for $8 each. Then they add the team's name to the T-shirts. The club sells the T-shirts for $15 each to raise money for the annual party. How much does the booster club make if they sell all the T-shirts?

Mark the Text

▢ Read the Problem and Search for Information · · · · · · · ·

Identify what information is given and what you are asked to find. Reread and mark the numbers and words that will help you.

▢ Decide What to Do and Use Your Ideas · · · · · · · · · · · · ·

You can use the strategy *Make a Table* to keep track of the details. Write what you know in the table. As you work through the steps below, complete the table.

Ask Yourself

If I know the number of T-shirts and the price of each, what operation can I use to find the total price?

T-shirts	Number	Price	Amount
Sold	75	$15	
Bought	75	$8	

Difference ⟶ ▭

Step 1 Find the total amount of sales.

Step 2 Find the total cost to buy the T-shirts.

Step 3 Find the difference between the total sales and the total cost to find how much the club makes.

So the booster club makes _____ .

▢ Review Your Work ·

Be sure to check your multiplication and subtraction.

Determine What tells you subtraction is needed to solve this problem?

_____

_____

Apply Your Skills

Solve the problems.

Ask Yourself

What number sentence can I use to find the amount of money the adult tickets raised?

② A school sells tickets to a basketball game to raise money. The tickets cost $12 for adults and $5 for students. Tickets are sold for 279 adults and 148 students. How much money is raised altogether?

Type of Ticket	Number of Tickets Sold	Price per Ticket	Amount
Adult			
Student			

◄ **Hint** First, estimate how much money each type of ticket raised. If your answer is close to the estimate, then your answer is reasonable.

Answer _____

Conclude Why is a table helpful for solving this problem?

_____

_____

③ Lori and Bill set up tables for the sports dinner. They used 26 rectangular tables that seat 12 people each. They also used 15 round tables that seat 8 people each. If 430 people come to the dinner, how many empty seats will there be?

Kind of Table	Number of Tables	Number of Seats per Table	Number of Seats
Rectangular			
Round			

Ask Yourself

Does the phrase *how many empty seats will there be* tell me to add or subtract?

Answer _____

Analyze Which operations did you use to solve the problem?

_____

_____

◄ **Hint** The question asks about empty seats, not how many seats in all.

Ask
Yourself

What information
tells you which
operations to use to
find the total points?

(4) Students play a game of tossing a soccer ball into a tub.
If the ball goes into the tub, they get 3 points. If the ball misses
the tub, they lose 2 points. Sam tosses the ball 20 times.
Sixteen of his throws go into the tub. The rest miss the tub.
How many points does Sam score altogether?

Hint Think about
how you will label
the columns.

Toss Result			
In Tub			
Misses Tub			

Answer _____

Contrast Suppose all of Sam's throws stay in the tub. Would you
use a table to solve this problem? Why or why not?

_____

_____

(5) A sports team holds two events. Tickets cost $3 for the
morning event. Tickets cost $4 for the afternoon event. A total
of 748 tickets are sold for the morning. There are 695 tickets
sold for the afternoon. How much more money is raised in the
afternoon than in the morning?

Hint Decide on the
steps you will use to
solve the problem.

Ask
Yourself

Will I add or subtract
to find the answer?

Answer _____

Judge What words in the problem tell you to compare? Are you
comparing tickets or money?

_____

_____

On Your Own

Solve the problems. Show your work.

6 The Central School District is adding soccer teams to 18 schools. The district has to buy 18 pairs of goals. Each pair of goals costs $498. The district also needs to buy 125 soccer balls for $19 each. How much will the district spend on these soccer supplies?

Answer _____

Revise What other question can you ask about the soccer supplies?

_____

_____

7 Teams ran in a relay race during a school field day. The school bought 15 each of three different trophies. First-place trophies cost $17 each. Second-place trophies cost $9 each. Third-place trophies cost $6 each. How much more did the school spend on first-place trophies than second-place trophies?

Answer _____

Propose What if the problem asked how much the trophies cost in all? Which operations would you use to solve it? Why?

_____

_____

Create Look back at Problem 4. Change the number of points for tosses that go into the tub and tosses that miss the tub. Write and solve a problem about a different student and a different number of tosses. Be sure your problem can be solved using the strategy *Make a Table*.

Strategy Focus
Use Logical Reasoning

MATH FOCUS: Division

Learn About It

Read the Problem

> Jen invites her friends to her house for a party. She tells them she lives in a red house. There is also a blue house and a green house on her street. The numbers of the three houses are 315, 451, and 550. Both 5 and 9 are factors of the number of the blue house. The green house's number is divisible by 5 and 11. What is Jen's house number?

Reread As you read the problem again, ask yourself questions.

- What is the problem about?

  _____

- What kinds of information are given?

  _____

- What am I asked to find?

  _____

Mark the Text

Search for Information

Read the problem again. Circle important numbers and words.

Record Write the information that you need.

The colors of the houses are _____ , _____ , _____ .

Jen lives in the _____ house.

The numbers of the houses are _____ , _____ , and _____ .

The number of the blue house has factors of _____ and _____ .

The number of the green house is divisible by _____ and _____ .

You can use this information to choose a problem-solving strategy.

Decide What to Do

You know the colors of the houses and the numbers.

Ask How can I find the number of Jen's house?

- I can use the strategy *Use Logical Reasoning*.

- I can use the clues and make a table to eliminate choices.

Use Your Ideas

Step 1 Use the first clue.

Both 5 and 9 are factors of the number of the blue house.

Five is not a factor of 451. Nine is not a factor of 550. Both 5 and 9 are factors of 315. So the number of the blue house is 315. No other house can be number 315.

	315	451	550
Red	X		
Blue	✔	X	X
Green	X		

Use an X when you eliminate a choice and a ✔ to show the right choice.

Step 2 Use the second clue.

The green house's number is divisible by 5 and 11.

550 is divisible by 5 and 11, so it could be the number of the green house. Check the other choice. 451 is not divisible by 5. So the green house must be number 550. No other house can be number 550.

The only remaining space shows the number of the red house. Put a ✔ in the space.

So Jen's house is number _____ .

Review Your Work

Reread the clues and check that your answers match.

Interpret In Step 2, why did you need to check if 451 was divisible by 5 and 11 after you had checked 550?

_____

_____

Try It

Solve the problem.

① A clue in a scavenger hunt says to be at the library at the mystery hour. To find the mystery hour, you need to find the missing digits in a division problem. The tens digit in the dividend is the same as the ones digit in the quotient. The sum of the digits of the quotient is 3 more than the divisor. The product of the digits of the quotient is 24. What is the mystery hour?

$$7 \overline{)\; 4 \; \square \; 8}$$

↑

The mystery hour is the tens digit of the dividend.

Mark the Text

▢ Read the Problem and Search for Information ⎫ . . . .

Notice that there is no remainder.

▢ Decide What to Do and Use Your Ideas ⎫ . . . .

You can use the strategy *Use Logical Reasoning*. Use what you know about the division words dividend, quotient, and divisor.

Step 1 Use the sum of the digits of the quotient.

3 more than the divisor → 7 + 3 = 10

The 10 possible quotients are:

_____

Ask Yourself

Could the quotient be as low as 19? Could it be as high as 91?

Step 2 Find which of the quotients have digits with a product of 24. The quotient could be _____ or _____ .

Step 3 Multiply by 7 to see if they match the dividend.

_____ × 7 = _____   _____ × 7 = _____

So the mystery hour is _____ o'clock.

▢ Review Your Work ⎫ . . . .

Make sure your dividend divided by 7 equals your quotient.

Conclude Is there any other possible dividend?

_____

_____

Apply Your Skills

Solve the problems.

2 Ken wrote this clue for a scavenger hunt: "Find the missing digits in the division problem. The dividend is a multiple of 9. The divisor is half the tens digit of the dividend. Write the three missing digits in order from least to greatest. This is the combination on a lock for the next clue." What is the combination?

$$\square \overline{)\ 2\ \square\ 1}\ \square\ 7$$

◀ **Hint** Find the quotient by dividing.

Ask Yourself

What do I know about multiples of 9?

The dividend is a multiple of 9, so it is _____ .

The divisor is half the tens digit of the dividend, so it is _____ .

Answer _____

Explain How can you use divisibility rules to check your answer?

_____

_____

3 Jon places a baseball, a basketball, and a football in front of three houses. The house numbers are 301, 305, and 308. He gives the following clues to his friend Jeff. What house number should Jeff go to if he wants the basketball?
- If you divide the house number with the baseball by 3, the remainder is 2.
- If you divide the house number with the basketball by 4, the remainder is 1.
- If you divide the house number with the football by 7, the remainder is 4.

◀ **Hint** Make a table to help eliminate choices.

Ask Yourself

If I divide 301 by 3, is the remainder 2?

	301	305	308
Baseball			
Basketball			
Football			

Answer _____

Examine A student says that the basketball is at house number 305. What error might the student have made?

_____

_____

4 Sarah stands near three doors. They are black, white, and red. She has three keys to unlock the doors. The keys are numbered 42, 56, and 63. Sarah has the following clues. Which key should she use to unlock the red door?
- The number of the key that unlocks the white door is even.
- The number of the key that unlocks the red door is divisible by 3.
- The number of the key that unlocks the black door is *not* divisible by 3.

Hint Write the numbers in the table. ▶

White			
Black			
Red			

Answer _____

Apply How did the table help you solve the problem?

_____

_____

Hint Remember the divisibility rules. ▶

5 Pia's special clue in her treasure hunt is to find the missing digits in the division problem. Together, the digits form a secret phone number. Write the complete quotient, dividend, and divisor one after the other. No two of the missing digits are the same and none are 0. The sum of the digits of the dividend is a multiple of 3. What is the secret phone number?

$$\begin{array}{r} 2\,●\,9 \\ ●\,\overline{)\,8\,●\,7} \end{array}$$

Ask Yourself

What number can be the missing digit in the dividend?

	Quotient		Dividend	Divisor
Phone number	2 ____ 9	–	8 ____ 7	____

Answer _____

Analyze Why is the clue that *no two missing digits are the same* important?

_____

_____

On Your Own

Solve the problems. Show your work.

6 Shana, Diana, and Elana each have a locker. Their lockers are the numbers 210, 660, and 990. One of them has a hidden prize in her locker. Shana's and Elana's locker numbers are divisible by 11. Shana's locker number is not divisible by 4. What is Shana's locker number?

Answer _____

Determine What information is given that is not needed to solve the problem?

_____

_____

7 There are two digits missing in this scavenger hunt clue. The number of seashells is a multiple of the number of friends. If you make a 2-digit number from the missing digits, you get a page number that will lead you to the next clue. The page number is not divisible by 7. What is the page number?

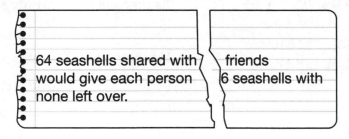

64 seashells shared with ⌐ friends
would give each person ⌐ 6 seashells with
none left over.

Answer _____

Infer How did you know that the number of friends was *not* a 2-digit number?

_____

_____

Look back at Problem 3. Change the addresses on the street and in the clues. Write and solve a problem about the location of one of the balls.

Create

In this unit, you worked with four problem-solving strategies. You can often use more than one strategy to solve a problem. So if a strategy does not seem to be working, try a different one.

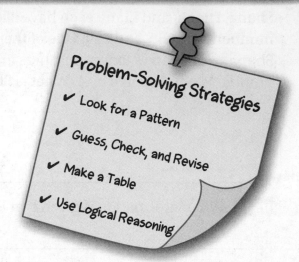

Problem-Solving Strategies
✔ Look for a Pattern
✔ Guess, Check, and Revise
✔ Make a Table
✔ Use Logical Reasoning

Solve each problem. Show your work. Record the strategy you use.

1. On Monday, the basketball team members sent 53 text messages. On Tuesday, they sent 45 text messages. On Wednesday, they sent 37 text messages. If the number of text messages the team members send changes by the same number each day, on what day will they send only 13 text messages?

2. Mr. and Mrs. Lopez and their 3 children pay $14 for tickets to a sporting event. Which game did they choose?

Basketball	Adults $5
	Children $2
Hockey	Adults $5
	Children $3
Football	Adults $4
	Children $2

Answer _____

Strategy _____

Answer _____

Strategy _____

3. Lia, Shayna, and Maggie each saved $144 to go to soccer camp. It took Lia 8 weeks to save her money. It took Shayna 4 weeks more than Lia to save her money. Maggie took twice as long as Lia to save her money. How much more per week did Shayna save than Maggie?

Answer _____

Strategy _____

4. Look at the following:

$$7 \times 6 \quad = 42$$
$$7 \times 66 \quad = 462$$
$$7 \times 666 \quad = 4,662$$
$$7 \times 6,666 = 46,662$$

Without multiplying, predict the product of $7 \times 66,666$.

Answer _____

Strategy _____

5. The local supermarket sells small oranges in crates that hold 18 oranges. It sells large oranges in crates that hold 24 oranges. In one week, it sells 126 crates of small oranges and 95 crates of large oranges. How many oranges does it sell?

Answer _____

Strategy _____

Explain how you know that your answer is reasonable.

Solve each problem. Show your work. Record the strategy you use.

6. Mr. Salvo rented a truck for 9 hours. He paid $19 an hour for the first four hours. He paid $17 an hour for the next two hours, and $15 an hour for the remaining time. What was the total cost for renting the truck?

    Answer _____

    Strategy _____

7. Sam is thinking of two numbers. The first number has one digit and the second number has three digits. The quotient of the two numbers is 75. Their product is 1,200. What are Sam's numbers?

    Answer _____

    Strategy _____

8. How many times will you write the digit 3 if you write all the whole numbers from 1 to 100?

    Answer _____

    Strategy _____

    Explain how you can find the answer without actually having to write all the numbers from 1 to 100.

9. The corner store sells lemonade in cans and bottles. Three cans and one bottle weigh the same as five cans. If one can weighs 16 ounces, what does one bottle weigh?

Answer _____

Strategy _____

10. Amy is 15 years younger than Alan. The sum of their ages is 101. How old are Amy and Alan?

Answer _____

Strategy _____

Write About It

Look back at Problem 4. Justify your prediction.

_____

_____

_____

Work Together: Buy Tickets

Your team is planning a trip to a new amusement park. Together, find the least expensive way to pay for rides. Should your team buy an *All Day Pass* or one or more *Ticket Books*?

Plan
1. Select which rides each of you will go on. You can choose a total of 10 rides each. You may choose the same ride more than once.

2. Determine how many tickets you need to buy.

3. Agree on the total number of tickets your team must buy.

Decide   Choose the option that is cheaper for your team.

Justify   Explain your team's decision. Use tables, lists, or other diagrams.

Present   As a group, share your decision with the class. Discuss your reasoning.

All-Day Pass $29
50-Ticket Book $38

Ride	Tickets
Bumper Cars	3
Carousel	2
Ferris Wheel	2
Gravitron	4
Log Flume	5
Roller Coaster	5
Scrambler	3
Sky Coaster	4

Unit Theme:
Cultures

You do not have to go far to learn about different cultures. Look around. You can learn about them right where you are! Try different foods and make arts and crafts. Take part in some celebrations. In this unit, you will see how math is important in every culture.

Math to Know

In this unit, you will use these math skills:

- Add and subtract like and unlike fractions
- Multiply and divide with fractions
- Add, subtract, multiply, and divide with decimals

Problem-Solving Strategies

- Draw a Diagram
- Solve a Simpler Problem
- Work Backward
- Guess, Check, and Revise

Link to the Theme

Write another paragraph about the results of Jared's poll. Include some of the facts from the table at the right.

There are 24 students in Jared's class. He takes a poll to see what languages they speak in addition to English. He shows his results to the class.

Language	Number of Students
Mandarin	3
Russian	1
Spanish	8

Use Math Language

Review Vocabulary

The list below shows vocabulary terms in this unit. Knowing the meaning of these terms will help you understand the problems.

decimal point   like denominators   numerator   unlike denominators
denominator   mixed number   simplest form   whole number

Vocabulary Activity   Word Roots

Word roots can help you understand the meaning of some math terms.

1. The Latin root *nominate* means *named*, or *called*. _____ has the same word root as *nominate*.

2. In the fraction $\frac{3}{5}$, the _____ names the fraction *fifths*.

3. *Numeral* has the same word root as _____ .

4. In the fraction $\frac{3}{5}$, the _____ tells the number of *fifths* you are talking about.

Graphic Organizer   Word Map

Complete the graphic organizer.

- Write your own definition of *mixed number*.

- Draw a picture or diagram to show what the term means.

- Write an example of a mixed number.

- Write a number that is *not* an example of a mixed number.

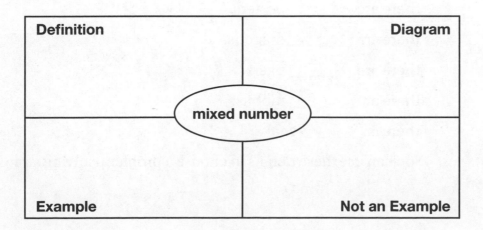

Strategy Focus
Draw a Diagram

MATH FOCUS: Addition and Subtraction with Fractions

Learn About It

■ Read the Problem ....

> Every year in the Chinese calendar is named after one of
> 12 animals. Eight students make posters showing the animals
> of their birth years. Three students draw dragons. Two students
> draw tigers. Two students draw rabbits. One student draws a
> snake. What fraction of all the posters show dragons or tigers?

Reread Think about what the problem tells you.

• What is the problem about?

_____

• How many students are there?

_____

• What do you need to find?

_____

Mark
the Text

■ Search for Information ....

Read the problem again. Circle the data and words that will help
you solve the problem.

Record Write the numbers for each type of animal poster.

There are _____ posters.

There are _____ dragons.

There are _____ tigers.

There are _____ rabbits.

There is _____ snake.

You can use these details to choose a problem-solving strategy.

■ Decide What to Do ⋯⋯⋯⋯⋯⋯⋯

You know there are 8 posters altogether. You also know the number of each kind of animal poster.

Ask How can I find the fraction that shows tigers or dragons?

- I can use the strategy *Draw a Diagram* to draw the 8 posters.

- I can find the fraction that are dragons and the fraction that are tigers. The numerator tells the number for a particular animal. The denominator tells the total number. Then I can use the diagram to find the sum of the fractions.

■ Use Your Ideas ⋯⋯⋯⋯⋯⋯⋯⋯

Complete the diagram as you work through the steps below.

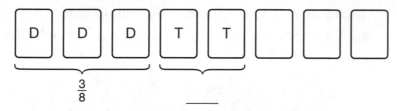

> The diagram can help you find the fraction of a whole. The number of Ds is the numerator of one fraction. The number of Ts is the numerator of the other.

Step 1 Draw and label a diagram to show the 8 posters. Use D for dragon, T for tiger, R for rabbit, and S for snake.

Step 2 Write fractions for the part of the posters that have a dragon and the part that have a tiger.

Step 3 Find the sum of the fractions.

How many posters are either dragons or tigers? _____

How many posters are there altogether? _____

$\frac{3}{8} + \frac{2}{8} =$ _____

The fraction of all the posters that show dragons or tigers is _____ .

■ Review Your Work ⋯⋯⋯⋯⋯⋯⋯

Check that your answer is in simplest form.

Describe How did making a diagram help you solve the problem?

_____

_____

Try It

Solve the problem.

① Maria and Carlo are making "worry dolls" to sell at the street fair. Maria has $3\frac{1}{8}$ balls of yarn. Carlo has $4\frac{1}{2}$ balls of yarn. They need a total of 8 balls for the dolls. Do they have enough?

Mark the Text

▨ Read the Problem and Search for Information ⌐ · · · · · · · ·

Identify the words and numbers you need to answer the question. Cross out any information you do not need.

▨ Decide What to Do and Use Your Ideas ⌐ · · · · · · · · · · · · ·

You can draw circles to represent the balls of yarn.

Step 1   Draw a diagram to represent the balls of yarn that Maria and Carlo have. Shade the balls and parts of balls that Maria and Carlo have.

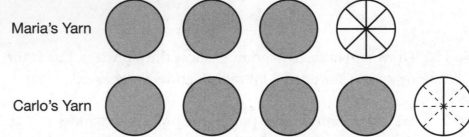

Maria's Yarn

Carlo's Yarn

Ask Yourself

> How many equal parts should I show in the circles?

Step 2   The fractions have unlike denominators. Write the fractions with like denominators.

How many eighths is $\frac{1}{2}$? $\frac{1}{2} = \dfrac{\square}{8}$

Find the total number of balls that Maria and Carlo have.

$3\frac{1}{8} + 4\frac{1}{2} = $ _____

They _____ have enough.

▨ Review Your Work ⌐ ·

Check that your diagrams show equal parts.

Explain   Why is $\frac{3}{8}$ *not* the answer to the question?

_____

_____

Apply Your Skills

Solve the problems.

(2) Kojo is weaving African cloth on a loom. He opens a package of 24 cloth strips. There are 6 red, 6 blue, 6 black, and 6 yellow strips. He uses 2 red, 4 blue, 3 black, and 5 yellow strips. What fraction of the strips did Kojo use? Write your answer in simplest form.

Ask Yourself

What does it mean for a fraction to be in simplest form?

Circle the strips that Kojo uses. Then add the fractions.

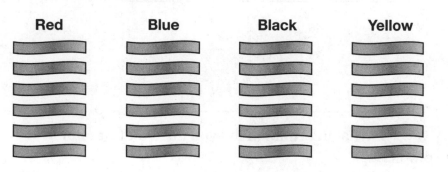

Red   Blue   Black   Yellow

◀ **Hint** Draw a diagram. Group the strips by color.

Answer _____

Interpret Patty says that Kojo used $\frac{5}{12}$ of the package. What mistake could Patty have made?

_____

_____

(3) Taj puts lanterns along a path that is $\frac{7}{10}$ kilometer long. Ravi puts lanterns along a path that is $\frac{3}{5}$ kilometer long. Whose path is longer? How much longer is it?

Shade the distance of each path.

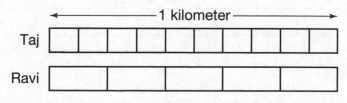

◀ **Hint** Make a diagram to compare the fractions.

Answer _____

Ask Yourself

How can my diagram show $\frac{7}{10}$ and $\frac{3}{5}$ with like denominators?

Contrast Suppose Taj puts lanterns along a path that is $\frac{5}{10}$ kilometer long. What would change in your diagram?

_____

_____

Hint Begin by finding how many yards the class needs in all.

▶

(4) An art class needs 2 feet of cloth to make each of 9 wall hangings. There are two rolls of cloth on the shelf. One roll is $3\frac{1}{3}$ yards long. The other is $2\frac{1}{2}$ yards long. Is there enough cloth for all the hangings? Explain.

←— 1 yard —→

Ask Yourself

How can I use fraction bars to show the length of cloth on each roll?

Answer _____

Determine What information did you need that was *not* in the problem?

_____

_____

Hint Start by changing the fraction parts of the mixed numbers so they have like denominators.

▶

(5) Mr. Ito folds paper to make animals. He has $3\frac{1}{2}$ pads of red paper, $2\frac{3}{5}$ pads of green paper, and $3\frac{1}{10}$ pads of yellow. The pads cost $7.50 each. Mr. Ito needs 9 pads of paper for a project. Will he have any paper left over? Explain.

←— 1 pad —→

Ask Yourself

Should I add the whole numbers or the fractions first?

Answer _____

Decide What information is *not* needed to solve the problem?

_____

_____

On Your Own

Solve the problems. Show your work.

(6) Kaya is making dream catchers. Each dream catcher uses 6 feathers. She makes 4 dream catchers with black feathers, 3 dream catchers with brown feathers, 1 dream catcher with white feathers, and 2 dream catchers with gray feathers. What fraction of the dream catchers have either black or brown feathers?

Answer _____

Relate What is another question you could ask about the dream catchers?

_____

_____

(7) Kim and Lee are making chains out of construction paper. Kim uses $2\frac{1}{2}$ sheets of red paper and $4\frac{1}{4}$ sheets of blue paper. Lee uses $3\frac{1}{4}$ sheets of red paper and $3\frac{3}{4}$ sheets of yellow paper. Who uses more paper? How much more?

Answer _____

Discuss Explain which operations you used to solve the problem.

_____

_____

Create Look back at Problem 2. Change the number of strips of cloth in the pack and the number of each color. Then change the number of strips that Kojo uses. Write and solve a new problem about the fraction of strips that are left.

Strategy Focus
Solve a Simpler Problem

MATH FOCUS: Multiplication and Division with Fractions

Learn About It

☐ Read the Problem ·····························

A recipe for a batch of 4 dozen teacakes is shown below. Tamara wants to make 12 dozen teacakes. She will sell them at the Food Fair. How many cups of pecans will Tamara need?

Teacakes (makes 4 dozen)
1 cup unsalted butter
$2\frac{1}{4}$ cups flour
1 teaspoon vanilla extract
$\frac{1}{2}$ cup confectioners' sugar
$\frac{1}{4}$ teaspoon salt
$\frac{3}{4}$ cup chopped pecans
$\frac{1}{2}$ cup candied cherries

Reread As you read the problem again, ask yourself these questions.

• What is the problem about?

_____

• What do I need to find?

_____

Mark the Text

☐ Search for Information ·····················

Read the problem again. Circle each number in the problem and in the recipe.

Record What information will help you solve this problem?

The recipe is for _____ teacakes.

The recipe calls for _____ cup chopped pecans.

Tamara wants to make _____ teacakes.

You can use these numbers to solve the problem.

Decide What to Do

You know that $\frac{3}{4}$ cup of pecans are needed to make 4 dozen teacakes and that Tamara wants to make 12 dozen teacakes.

Ask How can I find the number of cups of pecans needed for 12 dozen teacakes?

- I can use the strategy *Solve a Simpler Problem*.

- Substituting a whole number for the fraction $\frac{3}{4}$ will make the problem simpler.

- After I solve the problem using a whole number, I will understand how to solve it using the fraction.

Whole numbers are simpler to work with than fractions.

Use Your Ideas

Step 1 Tamara wants to make 12 dozen teacakes.

The recipe is for 4 dozen. 12 dozen = 3 × 4 dozen

Tamara wants to make _____ times as many teacakes.

Step 2 Set up and solve a simpler problem. Then solve the original problem.

Suppose the recipe called for 2 cups of pecans.

Simpler: 2 cups → _____ × 2 cups = _____ cups

Then follow the same procedure for $\frac{3}{4}$ cup of pecans.

Original: $\frac{3}{4}$ cup

_____ × $\frac{3}{4}$ cups → $\frac{3}{1} \times \frac{3}{4}$ = _____ cups

Write the answer as a mixed number. _____

Tamara needs _____ cups of pecans to make 12 dozen teacakes.

Review Your Work

Check that the product is less than the greater factor.

Explain How does thinking about simpler numbers help you?

_____

_____

Try It

Solve the problem.

(1) Jamal made 30 rolls for a party. He has 4 trays to put them on. His mom says that 30 rolls are not enough. How many pieces will Jamal have if he cuts each roll into halves? If he cuts each roll into fourths?

Mark the Text

▨ Read the Problem and Search for Information

Reread the problem and mark the numbers you will use to solve.

▨ Decide What to Do and Use Your Ideas

Use the strategy *Solve a Simpler Problem.*

Ask Yourself

How many pieces are made from one roll that is cut into halves? Into fourths?

Step 1 Set up and solve a simpler problem.
Simpler: Suppose Jamal made 5 rolls.
Mark the diagram to show each roll cut into halves.

$5 \div \frac{1}{2} \rightarrow \frac{5}{1} \times \frac{2}{1} = $ _____ halves.

Mark the diagram to show each roll cut into fourths.

$5 \div \frac{1}{4} \rightarrow \frac{5}{1} \times \frac{4}{1} = $ _____ fourths

Step 2 Follow the same procedure for the original problem.
Original: Jamal made 30 rolls.

_____ $\div \frac{1}{2} = $ _____ halves

_____ $\div \frac{1}{4} = $ _____ fourths

If Jamal cuts 30 rolls into halves, he will have _____ pieces.

If he cuts them into fourths, he will have _____ pieces.

▨ Review Your Work

Think about how your two answers are related.

Distinguish What information is *not* needed to solve the problem?

_____

_____

Apply Your Skills

Solve the problems.

(2) A recipe for chicken wings calls for 3 pounds of wings and $\frac{1}{4}$ cup chopped ginger. Tanya thinks 3 pounds of wings is more than she needs, so she decides to make half as much. How much chopped ginger should Tanya use if she makes half the recipe?

Ask Yourself

Would finding the amount of chopped ginger for four times the recipe be simpler?

Tanya needs _____ cup chopped ginger for the recipe.

Simpler: 4 times the recipe $4 \times \frac{1}{4}$ cup

Original: $\frac{1}{2}$ times the recipe   _____ $\times \frac{1}{4}$ cup

Hint Use the same steps as in the simpler problem.

Answer _____

Demonstrate How else could you use the strategy *Solve a Simpler Problem* to help solve this problem?

_____

_____

(3) Keisha is making her favorite lamb dish. To make 4 servings, the recipe calls for 1 cup of barley and $3\frac{1}{2}$ cups of boiling water. Ten people are coming for dinner. How much boiling water will Keisha need?

Hint Keisha needs to make more than twice as many servings.

Keisha needs _____ cups boiling water for 4 servings.

Simpler: 8 servings = 2×4 servings → 2 $\times$ _____ cups

Original: 10 servings = _____ $\times 4$ servings

→ _____ $\times$ _____ cups

Ask Yourself

How many times more than 4 is 10?

Answer _____

Determine What number sentences show the problems that you solved?

_____

_____

Hint Two operations are needed to answer this question.

(4) Anthony's pizza dough recipe calls for yeast, warm water, cooking oil, salt, and flour. He is making 8 pizzas. Anthony has 11 tablespoons of cooking oil. He needs $1\frac{1}{2}$ tablespoons of oil to make 1 pizza. Will 11 tablespoons of oil be enough? If not, how much more oil will he need?

Ask Yourself

If 1 tablespoon of oil is needed for each pizza, then how many tablespoons of oil does Anthony need?

Simpler: 1 tablespoon   8 × 1 tablespoon

Original: _____ tablespoons   8 × _____ tablespoons

Answer _____

Appraise Why is 1 tablespoon a good number to choose to make this a simpler problem?

_____

_____

(5) Every school week, cafeteria workers use $7\frac{1}{2}$ heads of lettuce for the salad bar. They use the same amount of lettuce each day. How many heads of lettuce are used each day?

Hint The problem may be easier to think about if you use 10 heads of lettuce.

Simpler: 10 heads of lettuce

$$\frac{\text{heads of lettuce}}{\text{days in a school week}} = \underline{\hspace{2cm}} \text{ heads of lettuce per day}$$

Ask Yourself

How many days are in a school week?

Answer _____

Relate What other question could you ask using the information in this problem?

_____

_____

On Your Own

Solve the problems. Show your work.

> Meat Pie (serves 6)
> 1 pound of lamb
> 2¼ cups chopped onions
> 1¾ cups chopped carrots

6 Here is a partial list of the ingredients for a meat pie. Peter wants to make enough pies to serve 18 at a party. How many cups of chopped onions will he need?

Answer _____

Analyze What important pieces of information are given in the recipe?

_____

_____

7 A chef has $4\frac{1}{2}$ gallons of olive oil. She uses $2\frac{1}{4}$ gallons of the oil. Then she pours equal amounts of the remaining oil into 3 bottles. How much oil does she pour into each bottle?

Answer _____

Choose What numbers are good to choose if you try solving a simpler problem first? Explain.

_____

_____

Create Use the recipe from Problem 6. Rewrite the amount of each ingredient so that your recipe serves either twice as many people or half as many people. Write and solve your new problem.

Strategy Focus
Work Backward

MATH FOCUS: Addition and Subtraction with Decimals

Learn About It

▧ Read the Problem ....

> The Romero family drove from Jacksonville, Florida to spend
> 3 days in Orlando. From there, they drove 235.1 miles to Miami.
> Then they drove 156.5 miles to visit cousins in Key West.
> Their total driving distance from Jacksonville to Key West was
> 543.5 miles. How far did they drive from Jacksonville to Orlando?

Reread Think about these questions as you read the problem.

• What is happening in this problem?

_____

• Which distances are given in the problem?

_____

• What question do I have to answer?

_____

Mark
the Text

▧ Search for Information ....

As you read the problem one more time, circle each distance.

Record Write the decimal that represents each distance.

It is _____ miles from Orlando to Miami.

It is _____ miles from Miami to Key West.

The Romero family drove _____ miles in all.

Think about how you could use this data to solve the problem.

Decide What to Do

You know two of the distances between the cities. You know the total distance.

Ask How can I find the distance from Jacksonville to Orlando?

- I can use the strategy *Work Backward*.

- I can start with the distance at the end of the trip. Then I can subtract the distances I know to find the missing distance.

Start with the total number of miles for the trip and work backward.

Use Your Ideas

Step 1 Write the total number of miles that the Romero family drove.

Step 2 Subtract the distance from Key West to Miami from the total distance driven. Remember to line up the numbers you are subtracting by their decimal points.

543.5 miles − _____ miles = _____ miles

Step 3 Now subtract the distance from Miami to Orlando.

387.0 miles − _____ miles = _____ miles

Your answer is the distance from Jacksonville to Orlando.

So the Romero family drove _____ miles from Jacksonville to Orlando.

Review Your Work

Use your answer to calculate the total distance driven. Check that this number matches the number in the problem.

Conclude Why is *Work Backward* a good strategy for this problem?

_____

_____

Try It

Solve the problem.

(1) Elsa is moving 5,000 miles away. She ships her bike and 2 boxes of equal weight to arrive 3 weeks ahead of her. Her bike weighs 48.5 pounds. The total shipment weighs 155.3 pounds. How much does each box weigh?

Mark the Text

▢ Read the Problem and Search for Information ⌐ · · · · · · · ·

Reread the problem and circle important numbers. Cross out any information you do not need.

▢ Decide What to Do and Use Your Ideas ⌐ · · · · · · · · · · · ·

You can use the strategy *Work Backward* to solve the problem.

Step 1   Start with the total weight. Subtract the weight of the bike to find the weight of the 2 boxes.

155.3 − _____ = _____ pounds

Step 2   Divide the weight by 2 to find the weight of each box.

_____ ÷ 2 = _____

So each box weighs _____ pounds.

Ask Yourself

After I find the weight of the 2 boxes together, how can I find what each one weighs?

▢ Review Your Work ⌐ ·

Check that the weights of the 2 boxes and the bike add up to 155.3 pounds.

Identify What information is given in the problem that is *not* needed to answer the question?

_____

_____

Apply Your Skills

Solve the problems.

(2) A customer came into Leo's fish shop and asked for two fish.
The first fish had a mass of 1.85 kilograms. The second fish
had a mass of 2.7 kilograms. The customer then asked for a
third fish. The total mass of the three fish was 5.64 kilograms.
What was the mass of the third fish?

◀ **Hint** You can work
backward to find the
mass of the third
fish.

Start with the total. The mass of all three fish together was
_____ kilograms.

The mass of the first fish was _____ kilograms.

The mass of the second fish was _____ kilograms.

Ask Yourself

What operation do
I need to do with
these numbers to
find the mass of the
third fish?

Answer _____

Explain Tell how you can check your answer.

_____

_____

(3) Sam travels to Tokyo, Japan. A tax of $20.00 is added to
the base price of the ticket. Then a $32.50 fee is added.
The total price of the ticket is $1,244.50. What is the base
price of the ticket?

◀ **Hint** Start with the
total price and work
backward.

Total Price	−	Fee	−	Tax	= Base Price
_____	−	_____	−	_____	= _____

Ask Yourself

In this problem,
should I add or
subtract when
working backward?

Answer _____

Generalize How does the word *added* help you know what to do
when working backward?

_____

_____

The distance I need
to find is in the
middle of the trip.
How can I still work
backward?

4 Lia and Joe went on an African safari. At 6.6 kilometers from camp, they saw some lions. They continued to travel, but stopped for water. From there, they traveled 14.3 kilometers and saw some zebras. At that point, they were 34.4 kilometers from where they started. How far did they travel between seeing the lions and stopping for water?

The total distance was _____ kilometers.

The distance they traveled before seeing the lions was _____ kilometers.

Hint Making a drawing to show the situation may help you solve this problem.

Answer _____

Determine How did you use the strategy *Work Backward* to solve the problem?

_____

_____

Hint Think of which numbers you need to add, subtract, and multiply.

5 Diego is going with his soccer team on a 3-day trip to the soccer playoffs. He will pay for part of his trip. His aunt gives him $50.00 for the trip and will pay for his room at a hotel. The cost of travel is $260.50. Diego needs $15.50 for each of the three days for food. How much does Diego need to pay for his trip?

The cost of travel plus food is _____ .

Ask
Yourself

Do I add or subtract
the cost of food?

Answer _____

Decide How can you estimate to see if your answer makes sense?

_____

_____

On Your Own

Solve the problems. Show your work.

6 Carrie is traveling with a red bag, a blue bag, and a black bag. She weighs them at the airport. The red bag weighs 16.93 pounds. The blue bag weighs 42.08 pounds. All three bags together weigh 92.6 pounds. How much does the black bag weigh?

Answer _____

Analyze How could you estimate the weight of the black bag?

_____

_____

7 Leng goes on a trip to Cambodia. The base cost of the plane ticket is $1,337.00. Taxes and fees of $100.68 are added to that cost. Leng also pays a fee for checking 2 bags. The fee is the same for each bag. The total cost of the trip is $1,538.68. How much does it cost to check one bag?

Answer _____

Assess Cathy says it costs $101.00 to check each bag. Is this correct? Explain.

_____

_____

Look back at Problem 5. Write a new problem by changing two of the numbers. Then solve your new problem.

Create

Strategy Focus
Guess, Check, and Revise

MATH FOCUS: Multiplication and Division with Decimals

Learn About It

☐ Read the Problem ....

> Paul is buying corn chips for a Cinco de Mayo festival. He buys 4 times as many pounds of white corn chips as blue corn chips. Paul bought 40 pounds of chips altogether. The chips cost $2.40 per pound. How much did Paul spend on white corn chips?

Reread As you read the problem, ask yourself these questions.

• What is the problem about?

_____

• What do you need to find?

_____

Mark the Text ✏

☐ Search for Information ....

Read the problem again. Circle important words and numbers.

Record Write the numbers you know.

Paul buys _____ times as many pounds of white corn chips as blue corn chips.

Paul bought _____ pounds of chips altogether.

The chips cost _____ per pound.

You can use this information to choose a problem-solving strategy.

Decide What to Do

You know the total number of pounds of chips. You know the relationship between the pounds of white chips and blue chips. You also know the cost of 1 pound of chips.

Ask How can I find the total cost of the white corn chips?

- I can use the strategy *Guess, Check, and Revise*.

- I can guess the number of pounds of blue corn chips. Then I can check if the total pounds of chips. I can make another guess.

- I can multiply by the cost per pound to find the total cost.

Use Your Ideas

Step 1   There are 40 pounds of chips in all. There are 4 times as many pounds of white chips as blue. Try 10 pounds of blue chips. Fill in the table as you complete the steps.

Pounds of Blue Chips	Pounds of White Chips	Total Pounds of Chips
10	4 × 10 = _____	10 + 40 = _____
5	4 × 5 = _____	5 + _____ = _____
8	4 × _____ = _____	8 + _____ = _____

> Use the results of your first guess to help make the second guess.

Step 2   50 is greater than 40. Try a lesser number of pounds of blue chips. Try 5 pounds of blue chips.

Step 3   25 is less than 40. Try a greater number of pounds of blue chips. Try 8 pounds of blue chips.

Step 4   8 pounds works! Now find the cost of the white chips.
$2.40 × _____ = _____

So Paul spent _____ on the white corn chips.

Review Your Work

Check that you answered the question that was asked.

Describe How did you use one guess to help make your next guess?

_____

_____

Try It

Solve the problem.

Mark the Text

Ask Yourself

How can I find how many boxes of each type of mooncake were sold?

1. A Chinese bakery sells red-bean mooncakes and lotus mooncakes during the Moon Festival. The bakery sold 5 more boxes of lotus mooncakes than red-bean mooncakes. Each box sells for $19.25. The bakery sold $750.75 worth of mooncakes. How many boxes of each type were sold?

▭ Read the Problem and Search for Information ┊ · · · · · · · ·

Reread the problem. Circle the important information.

▭ Decide What to Do and Use Your Ideas ┊ · · · · · · · · · · · · · ·

You can use the strategy *Guess, Check, and Revise* to find how many boxes of red-bean and lotus mooncakes were sold.

Step 1   Find the total number of boxes sold.

$750.75 ÷ $19.25 = _____

Step 2   Guess the number of boxes of red-bean mooncakes. Then find the number of boxes of lotus mooncakes. Check the total and revise your guesses as needed.

Red-Bean	Lotus	Total = 39?
15	15 + 5 = _____	15 + 20 = _____
20	20 + 5 = _____	20 + _____ = _____
17	17 + _____ = _____	17 + _____ = _____

So the bakery sold _____ boxes of red-bean mooncakes and _____ boxes of lotus mooncakes.

▭ Review Your Work ┊ ·

Multiply the number of boxes by the cost per box. Check that your answer is the total amount sold.

Explain   Why is a number between 15 and 20 a good third guess?

_____

_____

Apply Your Skills

Solve the problems.

(2) Kiran picks up Marcia on the way to a fireworks display. The distance from Marcia's house to the fireworks is half the distance from Kiran's house to Marcia's house. Altogether, Kiran drives 6.75 miles to get to the fireworks. How far is it from Kiran's house to Marcia's house?

Ask Yourself

If I guess the distance from Kiran's house to Marcia's house, how do I find the distance from Marcia's house to the fireworks?

Distance From Kiran's to Marcia's	Distance from Marcia's to Fireworks	Is the Total Distance 6.75 Miles?
5	$0.5 \times 5 =$ _____	$5 +$ _____ $=$ _____
4	$0.5 \times 4 =$ _____	
4.5		

Answer _____

Conclude How does making a table help with each guess?

_____

_____

Hint Your answer could be a decimal.

(3) A store clerk orders 12 boxes of silver candles and 12 boxes of gold candles. The gold candles cost $1.20 more per box than the silver candles. The total cost of the candles is $230.40. How much does each box of candles cost?

Ask Yourself

What would be a good first guess?

Cost of Silver Candles	Cost of Gold Candles	Is the Total Cost $230.40?
$10 \times 12 = $120	$11.20 \times 12 =$ _____	$120 +$ _____ $=$ _____
$8 \times 12 = $96	$9.20 \times 12 =$ _____	

Answer _____

Relate Mark first made a guess for the cost of the gold candles. How could he find the cost of the silver candles?

_____

_____

Hint Add $1.20 to the cost of the silver candles to find the cost of the gold candles.

4 Tanisha is setting up candles for a Kwanzaa celebration. Her sister Johanna is using strips of paper to make paper chains to decorate the house. She uses black, red, and green paper to make a total of 66 strips. Each strip is 6.75 inches long. She makes three times as many green strips as black strips. She makes half as many red strips as green strips. How many inches of each color paper does Johanna use?

Hint Guess the number of black strips. ▶

Black	Green	Red	Total
10			

Ask Yourself

Once I find the number of strips of each color, how do I find the length?

Answer _____

Determine What information is given that is *not* needed to answer the question?

_____

_____

Ask Yourself

Should I guess the number of small flags or large flags first?

5 Brett sold American flags at the Fourth of July parade. The prices are shown. He sold 3 times as many medium flags as large flags. He sold 5 times as many small flags as medium and large flags together. By the end of the day, he had sold a total of $874 in flags. What were Brett's sales in small flags?

Flags		
Small	4 in. × 6 in.	$1.95
Medium	2 ft × 3 ft	$15.25
Large	3 ft × 5 ft	$24.50

Large	Medium	Small	Total Sales

Hint Guess the number sold of each size flag. Then find the total sales with those numbers. ▶

Answer _____

Analyze When you *Guess, Check, and Revise*, is there only one way to make your guesses? Explain.

_____

_____

On Your Own

Solve the problems. Show your work.

6 The residents living on two streets bought trees to plant on Arbor Day. The residents of Elm Street bought 8 fewer trees than the residents of Oak Street. Altogether, they bought 24 trees. Each tree cost $29.99. How much did the residents of each street spend on trees?

Answer _____

Compare What is another way to compare the number of trees the residents of Elm Street bought to the number the residents of Oak Street bought?

_____

_____

7 At an autumn fair, there is a contest where growers win prizes for the heaviest pumpkins. The three heaviest pumpkins weigh a total of 809.5 pounds. The third-place pumpkin weighs 172 pounds. The second-place pumpkin weighs half as much as the first-place pumpkin. How much does the first-place pumpkin weigh?

Answer _____

Examine Dave thinks that the first-place pumpkin weighed 637.5 pounds. What mistake could Dave have made?

_____

_____

Create Look back at Problem 2. Write a new problem by changing the total distance to the fireworks and the relationship between the distance from Kiran's house to Marcia's and from Marcia's house to the fireworks. Solve your new problem using *Guess, Check, and Revise.*

In this unit, you worked with four problem-solving strategies. You can often use more than one strategy to solve a problem. So if a strategy does not seem to be working, try a different one.

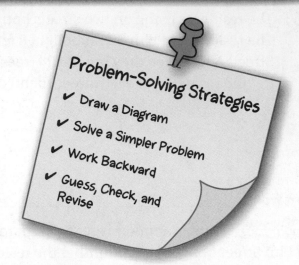

Problem-Solving Strategies

✔ Draw a Diagram

✔ Solve a Simpler Problem

✔ Work Backward

✔ Guess, Check, and Revise

Solve each problem. Show your work. Record the strategy you use.

1. Jenna thought of a decimal. She multiplied the decimal by 3. Then she multiplied the product by 5. She multiplied the next product by 2. Her final product was 375. What was her original decimal?

2. Duncan's recipe for cooking rice calls for 2 cups of water and 1 cup of rice. He can make 4 servings using that recipe. He wants to make 14 servings of rice. How much water should he use? How much rice?

Answer _____

Strategy _____

Answer _____

Strategy _____

3. Rico is making a budget for a 5-day vacation trip. According to his budget, he will spend $2\frac{1}{2}$ times the cost of his plane ticket for 5 nights at a hotel. He will spend $1\frac{1}{2}$ times the amount of his 5-day food budget on the plane ticket. If Rico has budgeted $36 *per day* for food, what is the cost of one night at the hotel?

Answer _____

Strategy _____

4. Trevor bought a basketball for his brother. The ball came in a cube-shaped box. Each face of the box is a $10\frac{3}{4}$ in. $\times$ $10\frac{3}{4}$ in. square. Trevor wraps ribbon around all sides of the box as shown. Then he uses 16 inches of ribbon to make a bow. How much ribbon does he use in all?

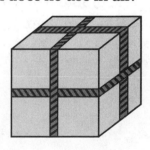

Answer _____

Strategy _____

5. Three florists sold roses for one hour on Mother's Day. Lily sold $\frac{1}{2}$ as many roses as Bud. Bud sold $\frac{1}{3}$ as many roses as Flora. Flora sold 144 roses. How many roses did Lily sell?

Answer _____

Strategy _____

Explain how you found your answer.

Solve each problem. Show your work. Record the strategy you use.

6. Phil has a recipe for pancakes that makes 4 servings. He only wants to make 2 servings. The recipe calls for $2\frac{3}{4}$ cups of milk. How much milk will Phil need?

Answer _____

Strategy _____

7. Donna paid $4.40 for a 2-mile taxi ride. The taxi company charges a certain amount for each $\frac{1}{5}$ mile traveled in the taxi. It charges twice that amount for the first $\frac{1}{5}$ mile traveled. How much does the taxi company charge for the first $\frac{1}{5}$ mile and each $\frac{1}{5}$ mile after the first one?

Answer _____

Strategy _____

8. Ms. Crane drives a school bus. She leaves her home and drives $1\frac{1}{2}$ miles north to pick up one student. Then she drives 6 miles east to pick up two more students. From there she drives $1\frac{1}{2}$ miles south to pick up the last student. Finally, she drives $3\frac{1}{2}$ miles west to the school. How far, and in which direction, is the school from her home?

Answer _____

Strategy _____

Explain how you could use the information in the problem to write an equation to solve the problem.

9. Ari bought four 10-ounce packages of strawberries at the farmers market. He also bought some $4\frac{1}{2}$-ounce packages of blueberries. Altogether, Ari bought $53\frac{1}{2}$ ounces of blueberries and strawberries. How many packages of blueberries did he buy?

Answer _____

Strategy _____

10. The distance around Cathy's rectangular garden is 37.2 meters. The length of her garden is twice its width. What are the measures of the length and width of Cathy's garden?

Answer _____

Strategy _____

Write About It

Look back at Problem 1. Tell what you know about the operations of multiplication and division that helped you solve this problem.

_____

_____

Work Together: Make Salsa

Your team is making salsa to sell at an International Food Festival. Together, you will determine how many jars to make and how to make the most money selling them.

Plan Find the amount of each ingredient you will need to make all the salsa. Find how much it costs to make each jar of salsa.

Decide Determine a reasonable price for which to sell your salsa. Estimate how many jars the team might sell and its total profit.

Create Make a poster to advertise your salsa. Include important information such as ingredients and the price of each jar. Will you offer any specials?

Present As a team, display your poster. Discuss how you decided upon the price, and how you found the total profit.

Salsa Recipe (makes 1 jar)
4 tomatoes
$\frac{1}{2}$ large onion
3 cloves of garlic
$\frac{1}{3}$ cup cilantro
1 jalapeño pepper
juice from $1\frac{1}{2}$ limes

Ingredient	Cost
1 tomato	$0.80
1 onion	$0.60
1 garlic clove	$0.05
1 cup cilantro	$1.50
1 jalapeño pepper	$0.45
1 lime	$0.50

Unit Theme:
Moving Along

Look around. There is always something on the move! On the road, you may see cars, bikes, or runners. Planes or blimps may fly overhead. Boats or surfers may skim along the water's surface. In this unit, you will see how math is part of how things move along.

Math to Know

In this unit, you will use these math skills:

- Write rates, ratios, and equivalent ratios

- Compute with fractions, decimals, and percents

- Apply number sense about percents

Problem-Solving Strategies

- Make a Table

- Write an Equation

- Use Logical Reasoning

Link to the Theme

Write another paragraph about the Dragon Boat Festival race. Include fractions, decimals, or percents in your paragraph.

Jamie is watching the annual Dragon Boat Festival race. There are 10 boats in the competition and 5 people on each team. So far, his brother's team is in the lead.

_____

_____

_____

_____

_____

_____

Use Math Language

Review Vocabulary

The list below shows vocabulary terms in this unit. Knowing the meaning of these terms will help you understand the problems.

decimal	equivalent fractions	fraction	rate
discount	equivalent ratios	percent	ratio

Vocabulary Activity   Math Words

Some terms are used only in math. Use terms from the list above to complete the following sentences.

1. Another way to say $\frac{1}{2}$ is 50 _____ .

2. In the number 2.5, the 5 comes after the _____ point.

3. If you have 2 apples and 6 oranges, the _____ of apples to oranges is 2:6.

4. $\frac{1}{3}$ and $\frac{2}{6}$ are _____ because 2:6 and 1:3 make the same comparisons.

Graphic Organizer   Word Map

Complete the graphic organizer.

- Write your own definition of *equivalent fractions*.

- Draw a diagram to show what *equivalent fractions* mean.

- Write an example of the term.

- Write something that is *not* an example of the term.

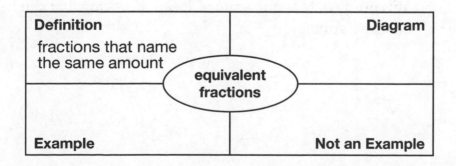

83

Strategy Focus
Make a Table

MATH FOCUS: Rates

Learn About It

▨ Read the Problem . . . .

> Nicole is skating to her friend's house. She skates 160 feet in 5 seconds. Her friend's house is 800 feet away. If Nicole keeps skating at that rate, how much time will it take her to skate to her friend's house?

Reread Ask yourself these questions as you read.

• What is the problem about?

_____

• What information is given?

_____

• What does the problem ask you to find?

_____

Mark
the Text

▨ Search for Information . . . .

Read the problem again. Look for important numbers and units.

Record Write what you know about the problem.

Nicole will skate _____ feet to reach her friend's house.

Nicole skates at a rate of _____ feet in _____ seconds.

This problem is about a rate of speed. Knowing that can help you choose a strategy.

Decide What to Do

You know how fast Nicole skates. You know how far she has to go.

Ask How can I find out how much time it will take Nicole to skate to her friend's house?

- I can use the strategy *Make a Table.*
- I can use the table to show how far Nicole will skate in 5 seconds, in 10 seconds, in 15 seconds, and so on.

Use Your Ideas

You can use a table to keep track of things that change, like distance and time.

Step 1 Make a table. Label your table.

Step 2 Fill in the first column to show how far Nicole skates in 5 seconds.

Distance Skated (feet)	160				
Time (seconds)	5				

Step 3 Fill in the table until you can see how much time it takes Nicole to skate 800 feet.

Distance Skated (feet)	160	320			
Time (seconds)	5	10	15	20	

It will take Nicole _____ seconds to skate to her friend's house.

Review Your Work

When you filled in the table you kept adding 160. You can multiply to check your addition.

Relate In any column of the table, if you divide the number of feet by the number of seconds, what will that quotient tell you?

_____

_____

Try It

Solve the problem.

1. A bus left for Parkville with 80 gallons of gas. Parkville is 350 miles away. After the bus went 50 miles it had used 6 gallons of gas. If the bus continues to use gas at the same rate, how much gas will be left when the bus gets to Parkville?

Mark the Text

▦ Read the Problem and Search for Information ·········

What does the problem ask you to find? Circle the important data.

▦ Decide What to Do and Use Your Ideas ···············

You can use the strategy *Make a Table* to keep track of the miles traveled and the gallons of gas left in the tank.

Fill in the table until you see how much gas it takes to go 350 miles.

Step 1   Make a table. Write labels to organize the table.

Step 2   Fill in the first column to show how much gas is left after 50 miles. Fill in the columns until you get to 350 miles.

> Ask Yourself
>
> What happens to the amount of gas in the tank as the bus travels every 50 miles?

Distance Traveled (miles)	50						
Gas Used (gallons)	6						
Gas Left (gallons)	74						

There will be _____ gallons of gas left when the bus gets to Parkville.

▦ Review Your Work ······························

Look at the numbers you used to fill in the table. Does your answer make sense?

Apply   How does the table help you solve the problem?

_____

_____

Apply Your Skills

Solve the problems.

(2) Some freight trains move at high rates of speed. The top speed of one train is 16 miles every 12 minutes. How long will it take that freight train, traveling at top speed, to travel 112 miles?

Ask Yourself

What numbers do I add to the distance and the time for each column?

Distance (miles)	16	32					
Time (minutes)	12	24					

Answer _____

Plan How did you know the operation to use to fill in the table?

Hint Estimate to see that your answer makes sense.

_____

_____

(3) A movie crew is filming people riding on sleds in the snow. Every 18 seconds the camera is running, it uses 24 feet of film. How many seconds can the movie crew shoot with a 120-foot roll of film?

Hint The rate is 24 feet of film every 18 seconds.

Film Used (feet)	24	48			
Time (seconds)	18	36			

Ask Yourself

How many seconds can be shot with 48 feet of film? With 72 feet of film?

Answer _____

Examine How did you know when you had found the answer to the problem?

_____

_____

Hint Be sure you
know the rate before
you begin filling in
the table.

(4) Some friends work at a local park on Cleanup Day. They have
cleaned an area of 500 square yards in 45 minutes. The friends
will work for a total of 3 hours. How many square yards will
the friends clean if they continue at the same rate?

Time (minutes)				
Area (square yards)				

Answer _____

Arrange How did you determine the numbers that belong in
the table?

_____

_____

Hint The bicycle
keeps moving
forward 30 yards
every time the wheel
makes 15 complete
turns.

Ask
Yourself

What labels do I
write in the first
column? What
numbers in the
second column?

(5) Berto is using his bicycle to measure the length of a bike trail.
Every time the wheel on his bike makes 15 complete turns, the
bicycle moves forward 30 yards. Berto counted 75 complete
turns. How far did the bicycle move as the wheel made
75 turns?

Answer _____

Explain How does showing a rate in a table help you to solve a
problem like this?

_____

_____

On Your Own

Solve the problems. Show your work.

6 Henry visits his family in San Francisco. They all ride a cable car through the city. It travels 350 feet in 25 seconds. How long does it take the cable car to travel 1,750 feet?

Answer _____

Identify What information in the problem is not needed to solve it?

_____

_____

7 Ms. Nelson is a truck driver. She left her home city with 260 gallons of fuel. Ms. Nelson's destination is 900 miles from her home city. She uses 20 gallons of fuel in the first 150 miles of her trip. The truck continues to use fuel at the same rate. How much fuel will Ms. Nelson have left when she reaches her destination?

Answer _____

Determine If you make a table to solve the problem, which labels will you use?

_____

_____

Create Look back at the problems in this lesson. Choose one problem and change at least two numbers to create a new problem. Solve your new problem.

Lesson 10

Strategy Focus
Write an Equation

MATH FOCUS: Equivalent Ratios

Learn About It

■ Read the Problem ....

> Two hours ago a helicopter took off with 200 gallons of fuel.
> It has used 50 gallons of fuel so far, but still needs to fly for
> another 4 hours. If the helicopter keeps using fuel at the same
> rate, how many gallons will be in the tank when it lands?

Reread Ask yourself these questions as you read.

- What is the problem about?

  _____

- What kinds of information are given?

  _____

- What does the problem ask you to find?

  _____

Mark
the Text

■ Search for Information ....

Read the problem again. Look for numbers that will help you solve
the problem.

Record Write the numbers that will help you solve the problem.

The helicopter started with _____ gallons of fuel.

It has been flying for _____ hours.

It has used _____ gallons of fuel.

It needs to fly for another _____ hours.

Think about how you can use this information to solve the problem.

Decide What to Do

You know how long the helicopter needs to fly and how much fuel it started with. You know the rate at which the helicopter uses fuel.

Ask How can I find the amount of fuel left in the tank?

- I can *Write an Equation* with words to show the relationship between the gallons of fuel used and the time in hours.

- I can fill in parts of the equation with the numbers I know, and solve the equation to find what is missing.

Use Your Ideas

Step 1 Write an equation in words with equivalent ratios.

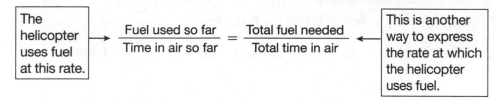

The helicopter has flown 2 hours and will fly 4 more hours. It will fly a total of 6 hours.

Step 2 Fill in the parts you know. Then find the missing parts.

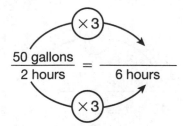

The helicopter will use a total of _____ gallons in all.

Step 3 Subtract the fuel used from the fuel at the beginning to find the amount of fuel left.

200 − _____ = _____

So the helicopter will have _____ gallons left when it lands.

Review Your Work

Check that your answer makes sense.

Explain How did equivalent ratios help you solve the problem?

_____

_____

Try It

Solve the problem.

(1) Marc is building a model airplane. The model is 8 inches long. The real airplane that the model is based on is 32 feet long. The wingspan of the model airplane is 10 inches long. How long is the wingspan of the real airplane?

 Mark the Text

▦ Read the Problem and Search for Information ⎸ · · · · · · · ·

What does the problem ask you to find? Circle the important data in the problem.

 **Ask Yourself**

How should I set up the equivalent ratios?

▦ Decide What to Do and Use Your Ideas ⎸ · · · · · · · · · · · · ·

You can use the strategy *Write an Equation.*

Step 1 Write an equation in words with equivalent ratios.

$$\frac{\text{model airplane length}}{\text{model airplane wingspan}} = \frac{\text{real airplane length}}{\text{real airplane wingspan}}$$

Step 2 Fill in the parts you know. Then find the missing part.

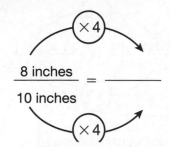

So the wingspan of the real airplane is _____ feet.

▦ Review Your Work ⎸ ·

Check that your ratios are equivalent.

(Interpret) How does multiplication help you solve the problem?

_____

_____

Apply Your Skills

Solve the problems.

(2) Sharon is building a model of a real glider. The model glider has a wingspan of 16 inches. The real glider has a wingspan of 48 feet. The length of the rudder on the model is 4 inches. What is the length of the real rudder?

$$\frac{\text{wingspan of model}}{\text{wingspan of real glider}} = \frac{\text{length of rudder on model}}{\text{length of real rudder}}$$

$$\frac{16 \text{ inches}}{48 \text{ feet}} = \underline{\hspace{2cm}}$$

Ask Yourself

What equation can I use to show how distances on the model are related to distances on the actual glider?

◀ **Hint** Write units carefully, because some distances are in feet and some are in inches.

Answer _____

Apply How could you have used a different ratio, such as *wingspan of model* to *length of rudder on model,* to solve the problem?

_____

_____

(3) A camera crew is filming an aerial action scene for a movie. The camera uses 3 feet of film every 2 seconds. How many seconds can the crew shoot with the 60 feet of film still on the roll?

$$\frac{\text{length of film}}{\text{shooting time}} = \frac{\text{length of film left}}{\text{shooting time left}}$$

$$\frac{3 \text{ feet}}{2 \text{ seconds}} = \underline{\hspace{2cm}}$$

Ask Yourself

What do I need to multiply by to find the shooting time left?

◀ **Hint** Every 2 seconds of shooting uses the same amount of film: 3 feet.

Answer _____

Demonstrate How could you use the same word equation to find how much film is needed to shoot for 2 minutes?

_____

_____

Ask Yourself

What should I multiply or divide?

Hint The ratio of any two distances on a scale drawing is the same as the ratio of those distances in real life.

▶

④ Jon is making a scale drawing of a jet that is 200 feet long. He wants the drawing to fit on a sheet of paper, so he makes the jet in the scale drawing 10 inches long. The tail of the actual jet is 60 feet tall. How tall should Jon make the tail on his drawing?

$$\frac{\text{length of jet on drawing}}{\text{height of tail on drawing}} = \underline{\hspace{4cm}}$$

$$\underline{\hspace{2cm}} = \underline{\hspace{1.5cm}}$$

Answer _____

Arrange How did you know where to put each phrase in your word equation?

_____

_____

⑤ A weather balloon can measure air temperatures as it ascends. A weather balloon rose 500 feet. At the same time, the temperature went down 2°F to 45°F. At that rate, what will the temperature be if the balloon rises another 2,500 feet?

$$\frac{\text{drop in temperature so far}}{\text{increase in height so far}} = \underline{\hspace{4cm}}$$

$$\underline{\hspace{2cm}} = \underline{\hspace{1.5cm}}$$

Ask Yourself

How can I find the change in temperature?

Hint The problem asks for the temperature, not just how much the temperature will drop.

▶

Answer _____

Identify How did you know which operations to use?

_____

_____

On Your Own

Solve the problems. Show your work.

(6) On the day of a large festival, a blimp has come to town. The blimp has just cruised 200 yards in 30 seconds. It continues at the same rate of speed. How long will it take the blimp to travel 1,800 more yards?

Answer _____

Formulate What is another question you could ask about the blimp?

_____

_____

(7) Ms. Powell is planning an air show for the Fourth of July. In one formation, planes A and B will fly 6 meters apart. She has drawn a diagram showing planes A and B. Where should Ms. Powell draw plane C to show that it will fly 42 meters to the right of plane B?

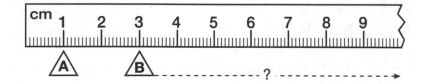

Answer _____

Justify What in this problem gives you a clue that you can use a strategy of writing an equation with equivalent ratios?

_____

_____

Choose one of the problems in this lesson. Change two or more of the numbers in the problem. Write and solve your problem.

Create

Strategy Focus
Use Logical Reasoning

MATH FOCUS: Comparing Fractions, Decimals, and Percents

Learn About It

▢ Read the Problem ....

> Murph's Surf Shop has 100 surfboards for renting. The boards
> are painted either red or blue. Each board has either stars or
> circles. Seventy-five of the boards are red. Fifty have stars, and
> 15% are blue with circles. How many red surfboards with stars
> are available to rent?

Reread Ask yourself these questions.

• What four types of surfboards does Murph's shop rent?

_____

• What is the problem asking?

_____

*Mark
the Text*

▢ Search for Information ....

Read the problem again. Circle the information about each type
of surfboard.

Record Write the information you know about the surfboards.

There are _____ surfboards in all.

_____ surfboards are red.

_____ surfboards have stars.

_____ of the surfboards are blue with circles.

You need a strategy that will help you find some hidden
information. Finding the hidden information will help you
solve the problem.

Decide What to Do

You know the fractions and percents given in the problem. You know that some surfboards belong to more than one clue.

Ask How can I find out how many red surfboards with stars are available to rent?

- I can *Use Logical Reasoning* to find hidden information in the information that is given.

Use Your Ideas

Make a table to keep track of your conclusions.

Step 1 Fill in the Total row.

There are 100 surfboards. Seventy-five are red, so _____ are blue.

	Red	Blue	Total
Stars			50
Circles			
Total	75		100

Step 2 Fill in the Total column.

Fifty surfboards have stars, so _____ surfboards have circles.

Step 3 Fill in the Blue column.

15% of the surfboards are blue with circles.
15% × 100 = _____

So _____ blue surfboards have stars.

$$15\% \times 100 = 0.15 \times 100$$

Step 4 Fill in the Red column.

Fifty surfboards have stars. Ten of them are blue.
So _____ of them are red.

Seventy-five surfboards are red. 40 of them have stars.
So _____ of them have circles.

So there are _____ red surfboards with stars to rent.

Review Your Work

Make sure the rows and columns add to the totals.

Describe How did the table help you use logical reasoning?

_____

_____

Try It

Solve the problem.

① *Daisy*, *Eli*, and *Froggy* are boats. One of them has $\frac{1}{4}$ tank of fuel. Another has 0.5 tank. The other has $\frac{3}{4}$ tank. Use the information shown to find how much fuel each boat has.

- *Daisy* has used at least 0.4 tank.
- *Froggy*'s tank is between 30% and 60% full.

Mark the Text

Read the Problem and Search for Information

You may want to find equivalent fractions for the decimals.

Decide What to Do and Use Your Ideas

You can use logical reasoning to eliminate possibilities.

Step 1 Use an X to mark what cannot be true. Use a ✔ to mark what is true.

Daisy has used at least 0.4 tank, so there is less than _____ tank left. So Daisy cannot have $\frac{3}{4}$ tank.

Froggy's tank is more than 30% and less than 60% full. So *Froggy* cannot have either $\frac{1}{4}$ or _____ tank.

Froggy has _____ tank.

Ask Yourself

How can information that rules out only one possibility be helpful?

	$\frac{1}{4}$ tank	0.5 tank	$\frac{3}{4}$ tank
Daisy			X
Eli			
Froggy	X		X

Step 2 Neither *Daisy* nor *Froggy* has $\frac{3}{4}$ tank. So _____ must have $\frac{3}{4}$ tank. *Daisy* must have _____ tank.

Daisy has _____ tank, *Eli* has _____ tank, and *Froggy* has _____ tank.

Review Your Work

Check that each row and column has two Xs and one ✔.

Conclude How did logical reasoning help you solve this problem?

_____

_____

Apply Your Skills

Solve the problems.

(2) Three jet skiers are in a race. Ben has just passed Alex at the $\frac{1}{4}$-mile mark. Carla is at the 0.3-mile mark. Who is leading? Who is in second place? Who is in third place?

	Leading	Second	Third
Alex		X	
Ben			X
Carla			

Ask Yourself

How can I eliminate some possibilities by using logical reasoning?

Answer _____

◀ **Hint** Try listing all the possible orders and crossing out the ones that do not fit the information in the problem.

Compare What other strategy might be used to solve this problem? Explain.

_____

_____

(3) Visitors at a lake can rent small or large boats. The lake has 50 boats, but 10 of them are broken. Half of the broken boats are small, but that is $\frac{1}{4}$ of all the small boats. How many of each size of boat are there? How many of each size are working?

	Large	Small	Total
Broken			10
Working			
Total			50

◀ **Hint** Use what you know to fill in part of the table.

Ask Yourself

How can I use logical reasoning to fill in more of the table?

Answer _____

Demonstrate Explain how you used logical reasoning to complete one column or row of the table above.

_____

_____

(4) A cruise ship has 1,000 cabins. One fourth of the cabins are inside and have no ocean views. The rest of the cabins are outside. One hundred fifty outside cabins are 3-room suites. There are 200 cabins in all that are 3-room suites. How many inside cabins are *not* 3-room suites?

	Standard Cabin	3-Room Suites	Total
Inside			
Outside		150	
Total			1,000

Answer _____

Analyze What other question could be answered using the information given in this problem?

_____

_____

Ask Yourself

What possibilities can I eliminate with the information in the first note?

(5) Workers are loading three cargo ships: the *Argo*, the *Bell*, and the *Crown*. One is 60% full. Another is 75% full. The third is 90% full. Use the notes to figure out which ship is which.

Bell has less than one third of its cargo left to load.

Argo has more than one third of its cargo left to load.

Bell is closer to being fully loaded than *Crown*.

	60% Full	75% Full	90% Full
Argo			
Bell			
Crown			

Answer _____

Identify What steps did you use to find the answer?

_____

_____

On Your Own

Solve the problems. Show your work.

(6) A local hobby shop has 80 boat kits in stock. Seventy-five percent of the kits are for sailboats. The rest are for canoes. Half the sailboat kits and one fourth of the canoe kits cost less than $30. How many of the kits cost more than $30?

Answer _____

Consider How would your answer change if there were 160 boat kits in all, but all the other numbers in the problem stayed the same?

_____

_____

(7) Three canoes are in a race. The red canoe is $\frac{7}{8}$ mile from the starting line and is not in second place. The brown canoe is 75% of the way to the 1-mile mark and is ahead of the green canoe. Which canoe is leading, which is second, and which is last?

Answer _____

Evaluate If any of the numbers in this problem were left out, could the problem still be solved?

_____

_____

Create

Choose one of the problems in this lesson. Change two or more of the numbers in the problem. Write and solve your problem.

Strategy Focus
Make a Table

MATH FOCUS: Percent of a Number and Other Percent Concepts

Learn About It

▢ Read the Problem ....

> A city is planning to host an annual marathon. This is the first year they will have one. They have limited the number of runners this year to 10,000. They expect to be able to increase the number of runners by 10% each year. At this rate, how many runners will be allowed to enter the marathon in the fourth year?

Reread Ask yourself questions as you read the problem again.

• How often will the race take place?

_____

• How does the city expect the number of runners to grow in the future?

_____

• What does the problem ask you to find?

_____

Mark the Text ✏ ╌╌▶

▢ Search for Information ....

Read the problem again. Circle important information about the race.

Record Write what you know about the race.

This year, the city will allow _____ runners to enter the marathon.

The second year, the race can have _____ percent more runners than the first.

The third year, the race can have _____ percent more runners than the second.

Notice that the number of runners will keep changing. This can help you choose a strategy.

Decide What to Do

You know the number of runners for the first year. You know that the city expects the number of runners to increase by the same percent for each year that follows.

Ask How can I find out how many runners will be allowed to enter the marathon in the fourth year?

- I can use the strategy *Make a Table*.

- I can find the number of new runners for each year and then find the total number of runners for each year.

Use Your Ideas

Step 1 Make a table. Show the year, the number of new runners allowed, and how many runners there can be in all.

Step 2 Compute and complete the table for the second, third, and fourth years.

Making a table is a good way to keep track of numbers that are changing.

Year	New Runners	Total Runners
1st	none	10,000
2nd	10% of 10,000 = 1,000	10,000 + 1,000 = 11,000
3rd	10% of 11,000 = _____	11,000 + _____ = _____
4th	_____ = _____	_____ = _____

So a total of _____ runners will be allowed to enter the race in the fourth year.

Review Your Work

Check that you found 10% of a different number of runners for each year.

Discover Why does the number of new runners keep changing even though the increase is always 10%?

_____

_____

103

Try It

Solve the problem.

① June is training for a marathon. She goes to a store that is having a sale on gear. All running shoes are 20% off. Everything else is 40% off. She buys a pair of running shoes that usually cost $50, a warm-up suit that usually costs $100, and 2 shirts that usually cost $25 each. How much does June spend, not including sales tax?

Mark the Text

▢ Read the Problem and Search for Information · · · · · · · ·

Reread the problem. Mark the important information.

▢ Decide What to Do and Use Your Ideas · · · · · · · · · · · · ·

Ask Yourself

How can I find the different sale prices and keep track of them?

You can use the strategy *Make a Table* to keep track of all the prices.

Step 1 Make a table to show each item, its regular price, the amount off, and its sale price.

Step 2 Complete the table.

Item	Regular Price	Amount Off	Sale Price
Shoes	$50	20% of $50 = $10	$50 − $10 = $40
Warm-up Suit	$100	40% of _____ = _____	$100 − _____ = _____
Shirts	2 for $50	_____ = _____	_____ = _____
		Total:	

So June spends _____ , not including sales tax.

▢ Review Your Work ·

Check that you used the correct percent with each item.

Assess How did making a table help you solve the problem?

_____

_____

Apply Your Skills

Solve the problems.

(2) Right now, Jon can ride his bike 50 miles in 3 hours 20 minutes. He wants to reduce his time by 10% each month. If he is successful, what will his time be in 2 months?

Ask Yourself

Will Jon's time increase or decrease each month?

	Change in Time (minutes)	New Time (minutes)
Now	none	200
1 month	10% of 200 = 20	200 − 20 = _____
2 months	10% of _____ = _____	_____ = _____

◀ **Hint** Rewrite 3 hours 20 minutes as 200 minutes.

Answer _____

Compare How is this problem similar to the Learn About It problem? How is it different?

_____

_____

(3) Mrs. Chen is buying skating equipment at a sale. Skates are 25% off. Helmets and knee pads are 60% off. She buys 2 pairs of skates that usually cost $100 each, 3 helmets that usually cost $50 each, and 3 sets of knee pads that usually cost $20 each. How much will Mrs. Chen spend before tax?

Item	Regular Price	Amount Off	
Skates	2 pairs for $200	25% of $200 = $ _____	$200 − $ _____ = $ _____
Helmets	3 for $150	60% of _____ = _____	_____ = _____
Pads		_____ = _____	_____ = _____
		Total before tax:	

◀ **Hint** Be sure to use the correct percent off for each item.

Ask Yourself

How should I label the last column of the table?

Answer _____

Extend If the sales tax is 5%, how would you find the final cost?

_____

_____

④ Tom wants to buy a go-cart for a race. The regular price is $200. It is on sale and has a red tag. How much will Tom save? How does that compare to 45% of the regular price?

> **25% off**
> Everything in the Store!
>
> Take an additional **20% off** any red-tag item!

Hint Compute the 25% discount first. ▶

Price	Discount	Dollars Off	Reduced Price
$200	25% off everything	25% of $200 = $50	$200 − $50 = _____
$150	20% off red-tag items	_____ = _____	_____ = _____

Answer _____

Probe Why do you think the amount Tom saves is different from taking the combined discounts from the regular price?

_____

_____

⑤ Mia wants to buy a model car for a race. The car she wants is $100. Because a new model is coming out soon, the store owner is reducing the price 10% each day until it sells. Mia only has $70. When will $70 be enough to buy the car?

Ask Yourself

Will my answer be an amount?

Hint Compare the new price for each day to the amount Mia has. ▶

Days	Discount	New Price
0	none	$100
1	10% of $100 = $10	$100 − $10 = $90
2		
3		
4		

Answer _____

Elaborate Suppose the owner reduced the price by $10 each day instead of by 10%. Would you still make a table to solve the problem? Why or why not?

_____

_____

On Your Own

Solve the problems. Show your work.

6 Tickets to a car race cost $15 for adults and $9 for children under 12. Groups of 10 or more get 20% off the total cost of their tickets. How much would tickets cost for a group of 4 adults and 10 children under 12?

Answer _____

Investigate What types of information did you organize to help you solve the problem?

_____

_____

7 Fifty children competed in a sack race at the city park. The number of racers has been increasing at a rate of about 20% each year. If that rate continues, would you expect there to be more than 90 racers participating in the sack race in 3 years? Explain.

Answer _____

Interpret Why is the exact number of racers you found *not* the number you should use to help you solve this problem?

_____

_____

 Choose one of the problems in this lesson. Change two or more of the numbers in the problem. Write and solve your new problem.

Create

In this unit, you worked with three problem-solving strategies. You can often use more than one strategy to solve a problem. So if a strategy does not seem to be working, try a different one.

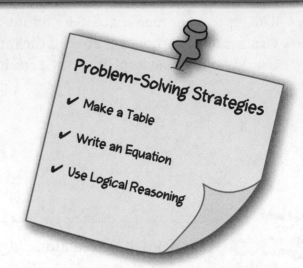

Problem-Solving Strategies

✔ Make a Table

✔ Write an Equation

✔ Use Logical Reasoning

Solve each problem. Show your work. Record the strategy you use.

1. Students at Oakmont School are going on a field trip. There must be at least 2 adults for every 15 students on the trip. So far, 75 students have signed up for the trip. How many adults must go with them?

2. Juana is hosting a party next month. She started to write her invitations at 1:15 P.M. Twenty minutes later, she had written 6 of the 24 she needs. At that rate, when will Juana finish?

Answer _____

Strategy _____

Answer _____

Strategy _____

3. Ahmed plans to buy art supplies on sale. At a $\frac{1}{3}$-off sale, he will buy an easel that usually costs $99. At a 20%-off sale, he will buy two sketchbooks and a set of brushes. The sketchbooks are usually $15 each and the set of brushes usually sells for $20. How much will Ahmed pay for all these supplies before taxes?

Answer _____

Strategy _____

4. Ruth, Sally, and Tasha took a Culture Quiz with 60 questions. Their scores were 75%, 80%, and 90%. Ruth got more than 45 questions right. Sally missed more than 10 questions. Tasha got between 10 and 14 questions wrong. Which student had which score?

Answer _____

Strategy _____

5. Mina is studying insects for her science project. She watches and times one insect as it crawls across her patio. It travels 2 feet 5 inches in 15 seconds. At that rate, how far would it crawl in one minute?

Answer _____

Strategy _____

Explain how you could solve the problem using a different strategy. You do not have to solve it again, just describe what you would do.

Solve each problem. Show your work. Record the strategy you use.

6. The library received 60 new books. Two-thirds are fiction and $\frac{2}{3}$ are paperback. Ten percent are hardbound, nonfiction books. Each book is either fiction or nonfiction and hardbound or paperback. How many of each type of book did the library receive?

Answer _____

Strategy _____

7. On a city map, 2 centimeters represents 5 kilometers. The distance from the bus station to the airport is 14 centimeters on the map. What is the actual distance between the bus station and the airport?

Answer _____

Strategy _____

8. Three stores are having different sales. One store is giving $\frac{1}{3}$ off, one is giving 40% off, and another is giving $10 off. A customer saved $25 buying a tent at store A. At store B, a $30 backpack was on sale for $20. A $60 camping stove cost the least at store C. Which store had which sale?

Answer _____

Strategy _____

Show how your answer can fit all the information given in the problem.

9. In its first year, a company has 80 workers. It plans to increase that number by 50 percent each year. If it does, how many workers will the company have in its fourth year?

Answer _____

Strategy _____

10. A recipe for 8 nut squares calls for 3 tablespoons of butter. Jon wants to make 4 dozen nut squares. How much butter does he need?

Answer _____

Strategy _____

Write About It

Problems 3 and 8 both involve different sales. Did you use the same strategy for both problems? Explain why or why not.

_____

_____

_____

Work Together: Plan a Trip

Your group is planning to buy equipment and supplies for a camping trip. Together with the other members of your group, find the least expensive way to buy your items. You do not need to buy everything at the same store.

Plan
1. List the equipment and supplies you will buy.
2. Decide upon a reasonable regular price for each item you listed.

Decide Choose where you will buy each item so that you spend as little as possible for everything you listed.

Justify Explain your group's decisions. Use tables, lists, or other diagrams.

Present As a group, share your decisions with the class. Discuss your reasoning.

Things to Bring
- Pillow
- Tent
- Toothbrush
- Toothpaste
- Soap
- Sleeping Bag
- Camp Stove
- Flashlight
- Batteries
- Tarp

CAMP STUFF
Buy 1
Get 2nd for
50% off!

CAMPING WORLD
Buy 2 items,
Get a 3rd for **FREE**

Rough-It
20% off **ALL** items!
Take an **extra 10%**
off if you pay with a
Rough-It charge card

Unit Theme:
Community

When you look around your community, what do you see? You may see buildings of all shapes and sizes. You may also see neighbors planting fruits and vegetables in a community garden. There could be a recycling center or an art museum around the corner. In this unit, you will see how math plays a role in every community.

Math to Know

In this unit, you will use these math skills:

- Use properties of two- and three-dimensional shapes

- Use formulas to find perimeter, area, circumference, and volume

- Identify line symmetry and rotational symmetry

Problem-Solving Strategies

- Work Backward

- Write an Equation

- Solve a Simpler Problem

- Draw a Diagram

Link to the Theme

Write another paragraph about Nisa's drawing. Include words that describe size and shape in your paragraph.

Nisa's dad is designing a new town hall. She wants to help. She shows her drawing to her dad and tells him about it.

_____

_____

_____

_____

_____

_____

Use Math Language

Review Vocabulary

The list below shows vocabulary terms in this unit. Knowing the meaning of these terms will help you understand the problems.

area	diameter	parallel	perpendicular
circumference	intersect	perimeter	radius

Vocabulary Activity   Word Pairs

Math terms are often learned together, but mean different things. Use terms from the above list to complete the following sentences.

Distance Around

1. Ben walks around the edge of a circular fountain. The distance around the circular fountain is called the _____ .

2. Then Ben walked around the edge of a rectangular pool. The distance around the rectangular pool is called the _____ .

Properties of a Circle

3. Ben sees a circular clock on a building. The distance from the rim above the 12 to the rim below the 6 through the center is called the _____ of the circle.

4. The distance from the rim above the 12 to the center of the circle is called the _____ of the circle.

Graphic Organizer   Word Circle

Complete the graphic organizer.

- Cross out the word that does *not* belong.

- Explain what the remaining three words have in common.

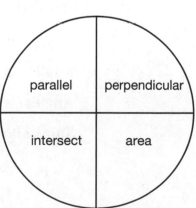

_____

_____

MATH FOCUS: Lines, Angles, Triangles, and Quadrilaterals

Learn About It

▢ Read the Problem ·

> Tyler draws a map of the community garden. The garden is in the shape of a trapezoid. He starts at one corner and measures each angle. He found that angle A is 3 times the measure of angle B. Angle C is 2 times the measure of angle B. Angle C is a right angle. What is the measure of angle D?

Reread Ask yourself these questions as you read the problem again.

• What shape is the garden Tyler draws?

_____

• What do you need to find?

_____

Mark the Text ✏

▢ Search for Information ·

Read the problem again and underline the words and numbers that will help you solve the problem.

Record Write what you know about the angles.

Angle C is a _____ angle.

Angle C is _____ times the measure of angle B.

Angle A is _____ times the measure of angle B.

Think about how you can use this information to solve the problem.

Decide What to Do

You know the garden is shaped like a trapezoid. You know how the measures of three of the angles relate to each other.

Ask How can I find the measure of angle D?

- I can start with what I know about the angles in a trapezoid and use the strategy *Work Backward*.

- I can find the measures of angles A, B, and C.

- I can subtract the measures of the angles I know from 360° to find the measure of angle D.

Use Your Ideas

The sum of the measures of the angles in a quadrilateral is 360°.

Step 1 Find the measures of angles A, B, and C.

The measure of angle C is _____ .

That is 2 times the measure of angle B.

So angle B measures _____ .

The measure of angle A is 3 times the measure of angle B.

So angle A measures _____ .

Step 2 Subtract the angle measures you know from 360.

$360 - 90 -$ _____ $-$ _____ $=$ _____

So the measure of angle D is _____ .

Review Your Work

Check your work by drawing and labeling the trapezoid and adding the four angle measures.

Compare Why is working backward a good strategy for this problem?

_____

_____

Try It

Solve the problem.

1. Dana left her hat at the bookstore and needs to walk back to get it. She is on A Street. Earlier, she bought a snack on a street parallel to A Street. Before that, she was at the bookstore on a street that intersects both A Street and the street where she bought the snack. On which street is the bookstore?

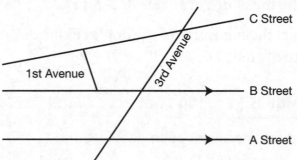

Mark the Text

▣ Read the Problem and Search for Information ·········

Circle words and phrases that will help you with the solution.

▣ Decide What to Do and Use Your Ideas ···············

You can use the *Work Backward* strategy.

Step 1 Mark the map with a *D* on _____ to show where Dana is now.

> **Ask Yourself**
>
> Am I looking for where Dana started her trip or where she finished her trip?

Step 2 The street where Dana bought the book is parallel to the street where you put the *D*. Put an *S* on _____, the street where she bought the snack.

Step 3 The bookstore is on a street that intersects A Street and the street where you put the *S*.

So Dana left her hat at the bookstore on _____ .

▣ Review Your Work ······························

Reread the problem to be sure you followed Dana's path correctly.

Examine There are two streets that intersect the street where Dana bought the snack. How do you know where she left her hat?

_____

_____

Apply Your Skills

Solve the problems.

(2) Lisa is drawing a map of a garden. The garden is in the shape of a scalene right triangle. The largest angle is 6 times the measure of the smallest angle. What is the measure of the smallest angle in the garden?

$$180 - 90 = \underline{\hspace{2cm}}$$

$$90 \div \underline{\hspace{2cm}} = \underline{\hspace{2cm}}$$

$$90 - \underline{\hspace{2cm}} = \underline{\hspace{2cm}}$$

◀ **Hint** No angles are equal in measure in a scalene triangle.

Ask Yourself

What is the measure of the largest angle in a right triangle?

Answer _____

Identify What phrase in the problem tells you division is one of the operations to use to solve it? How do you know?

_____

_____

(3) Alex and José walked down the street where José lives and turned onto a street perpendicular to the street José lives on. They walked down this street past one perpendicular street and to the next intersection. They were at the corner of Blake Street and 20th Street. What street does José live on?

22nd Street
21st Street
20th Street
Blake Street
Market Street

◀ **Hint** Start at Blake Street and 20th Street. Find the direction you can walk so that you pass one perpendicular street.

Ask Yourself

Were they walking down Blake Street or 20th Street when they finished?

Retrace Alex's and José's steps. Start from _____ Street and _____ Street and go past one perpendicular street before turning on another perpendicular street.

Answer _____

Conclude How does using the map help you to work backward?

_____

_____

4 Mr. Kee is drawing a map. So far, he has written only one street name—Evergreen Avenue. North Avenue intersects Ocean Avenue and is perpendicular to Damen Avenue. North Avenue is parallel to Lake Street. Evergreen Avenue intersects Blue Avenue and Lake Street at the same spot. Which lettered road is Damen Avenue?

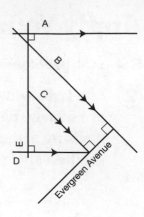

Hint Write the names of the other streets on the map as you determine them.

Ask Yourself

Where does Evergreen Avenue intersect two other roads at the same point?

Answer _____

Explain Do you need to know that Ocean Avenue and Blue Avenue are parallel to solve the problem? Why or why not?

_____

_____

5 Sandy is drawing a map of her garden. She could not measure angle C.
The difference between the measures of angle B and angle A is 29°. Angle D is 11° more than angle B. Angle D measures 120°. What is the measure of angle C?

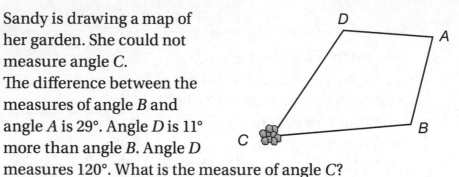

Hint Start with the only angle measure you know.

Angle D = 120°

Angle B = 120°   − 11° = _____ °

Angle A = _____ ° − 29° = _____ °

Ask Yourself

Is the sum of the angle measures the same for all four-sided figures?

Answer _____

Determine How can I use addition to check my answer correct?

_____

_____

On Your Own

Solve the problems. Show your work.

(6) Perry is looking at map as he walks through his new neighborhood. He is on Elm Street. Earlier, he crossed a street that is parallel to Elm Street. The street that intersects the two parallel streets is perpendicular to the street he lives on. What street does Perry live on?

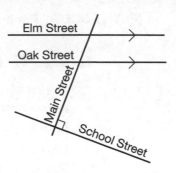

Answer _____

Judge How do you know the other streets *cannot* be where he lives?

_____

_____

(7) Joel is sketching a floor plan for a community center. Angle *A* is 10° greater than angle *B*. Angle *B* measures 15° more than twice the measure of angle *C*. Angle *C* is 55°. What is the measure of angle *D*?

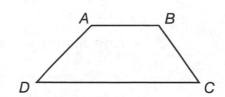

Answer _____

Decide Which angle do you solve for first? Explain.

_____

_____

Write a problem that can be solved using the strategy *Work Backward*. Write about a triangular map. Your problem should ask for one angle measure. Solve your problem.

Create

Strategy Focus
Write an Equation

MATH FOCUS: Perimeter, Area, Circumference

Learn About It

▊ Read the Problem ....

> Corrie buys special cloth that helps keep weeds out of the community garden. Ten rectangular plots in the garden are 4 feet by 6 feet. Four rectangular plots are 10 feet by 10 feet. The cloth costs $1 for each square foot. How much does Corrie spend on cloth for all 14 plots?

Reread As you read the problem, ask yourself these questions.

• What is Corrie doing?

_____

• What shape is each plot in the community garden?

_____

• How many plots are there in the community garden?

_____

• What am I being asked to find?

_____

Mark the Text →

▊ Search for Information ....

Read the problem again. Circle the information about the sizes of the garden plots.

Record Write what you know about each plot in the garden.

_____ plots are 4 feet by 6 feet.

_____ plots are 10 feet by 10 feet.

The cloth costs _____ .

Think about how you can use this information to solve the problem.

Decide What to Do

You know the dimensions of the plots in the garden. You know how many plots there are of each size. You know the cost of 1 square foot of cloth.

Ask How can I find the cost of the cloth?

- I can use the strategy *Write an Equation*.

- First, I can write an equation to find the area of each plot. Next, I can find the total area of the plots. Then I can find the cost to cover all of the plots with the cloth.

Use Your Ideas

Step 1 Write an equation to find the area of one of each size plot.

4 feet × 6 feet = _____ square feet

10 feet × 10 feet = _____ square feet

Step 2 Multiply the areas of the plots by the number of each size. Then add to find the total area.

_____ × 24 square feet = 240 square feet

_____ × 100 square feet = 400 square feet

240 square feet + 400 square feet = _____

Step 3 Multiply the total area by the cost of 1 square foot of cloth.

640 × _____ = _____

So the total cost of the cloth for the garden is _____ .

> Area of a rectangle = length × width

Review Your Work

Check that you used the correct length and width in each equation.

`Describe` How could you write one equation to find the area of all ten garden plots that are 4 feet by 6 feet?

_____

_____

Try It

Solve the problem.

① Taku's garden is a rectangle with a semicircle on each end. The semicircle has a diameter of 2 meters. He puts a border around his garden. The border comes in pieces that are each 1 meter in length. How many pieces of border will Taku need?

2 m ← 3 m →

Mark the Text

▢ Read the Problem and Search for Information ········

Mark details that will help you to find a solution.

▢ Decide What to Do and Use Your Ideas ·············

Write an equation to help you find the perimeter of the garden.

Perimeter = length of 2 curved sides + length of 2 straight sides

Step 1 List the parts of the garden that will need pieces of border.

- The straight sides are each _____ meters long.

- Each semicircle has a diameter of _____ meters.

Step 2 The 2 semicircles have the same circumference as a circle with a diameter of 2 meters. Use 3.14 for π.

Circumference = 2 × π × radius

= 2 × π × _____

The circumference is about _____ meters.

The length of 2 long sides of the rectangle is _____ meters.

The perimeter is about _____ + _____ meters.

Taku will need _____ pieces of garden border.

Ask Yourself

How do the lengths of a diameter and a radius compare?

▢ Review Your Work ·····························

Did you add all of the parts that need pieces of border?

Interpret Why is your answer different than the perimeter?

_____

_____

Apply Your Skills

Solve the problems.

(2) Tim and his neighbors need to put up fencing around each of their gardens to keep the rabbits out. They will buy a length of 1-meter tall fencing to share. All of the gardens are rectangles. Three gardens measure 4 meters by 8 meters. Four gardens measure 8 meters by 6 meters. How much fencing will the neighbors need?

Ask Yourself

Do I need to find the perimeters or the areas of the gardens?

Perimeter of 4 meter by 8 meter gardens:

2 × _____ meters + 2 × _____ meters = _____ meters

◀ **Hint** Perimeter of Rectangle = $2l + 2w$.

Perimeter of 8 meter by 6 meter gardens:

2 × _____ meters + 2 × _____ meters = _____ meters

Total length of fencing needed:

3 × _____ meters + 4 × _____ meters = _____ meters

Answer _____

Explain How do you know which operations to use to solve this problem?

_____

_____

(3) Manny is planting a garden. He will plant tomatoes, peppers, and squash. He uses a corner of the backyard that is shaped like a right triangle. The legs of the triangle measure 12 feet and 27 feet. One bag of fertilizer covers 100 square feet. How many bags of fertilizer does Manny need to cover the entire garden?

◀ **Hint** Draw a diagram of the garden to help you find the base and height.

Area of a triangle = $\frac{1}{2}$ × base × height

$\frac{1}{2}$ × _____ × _____ = _____

Ask Yourself

Which sides are the base and the height of a right triangle?

Answer _____

Identify What information is given that is *not* needed?

_____

_____

4 David is making circular flowerbeds in front of the library.
He needs to lay out the circles with twine before planting.
He will have one circle with a 2-meter radius. He will have
three smaller circles each with a 1-meter radius. He has
25 meters of twine. Will David need more twine to lay out
his garden? Explain.

Hint You can use
3.14 as the value of π.

▶ Circumference of a circle = 2 × π × _____

Answer _____

Determine What single equation could you use to find the amount
of twine needed?

_____

_____

5 The walkway in a community garden is made up of these
congruent right triangles. Ben paints each triangle a different
color. He alternates between two colors. One can of paint
covers 2 square yards. How many cans of paint of each color
does Ben need?

Ask
Yourself

How many triangles
of each color
are there?

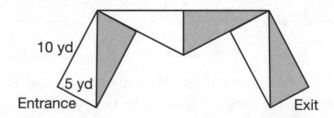

10 yd

5 yd

Entrance   Exit

Area of one triangle:
$\frac{1}{2}$ × _____ yards × _____ yards = _____ square yards

Hint Ben cannot
buy part of a can of
paint.

▶ **Answer** _____

Examine One student finds that the area of each section is
25 square yards. The student says that since 13 cans are needed to
cover one section, 39 are needed to cover three sections.
Is this correct? Why or why not?

_____

_____

On Your Own

Solve the problems. Show your work.

(6) Madison made four square wooden frames for her garden. The frames measure 7 feet on a side. She is making plastic covers for the frames to protect the plants in case of a late frost. Each frame needs a square piece of plastic large enough to cover the frame and hang 1 foot over on each side. How much plastic does Madison need?

Answer _____

Design Draw a diagram to show how you found the length of a side of each sheet of plastic.

(7) Delia designed her flower garden. Her design uses squares and equilateral triangles. Each shape is 24 inches on a side. She will put a 25-inch tall fence around the perimeter of the garden. What total length of fence will Delia need?

Answer _____

Conclude Why would using formulas for finding the perimeter of a square or of a triangle *not* help you solve this problem?

_____

_____

Create Write a problem that can be solved using the strategy *Write an Equation*. The problem should be about putting fence around a garden that is made up of a rectangle and a triangle. Solve your problem.

Strategy Focus
Solve a Simpler Problem

MATH FOCUS: Solid Figures and Volume

Learn About It

Read the Problem

There is still some waste left even after things are recycled. At a local recycling center, there is a bin used to collect the waste. After it is collected, it is compacted and taken away. The bin is shown below. What is the volume of this bin?

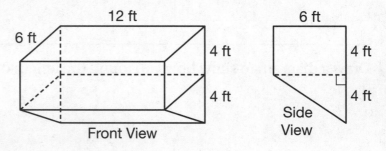

Front View · Side View

Reread Ask yourself these questions as you read the problem.

• What kind of information is given?

_____

• What am I asked to find?

_____

Mark the Text

Search for Information

Read the problem and study the diagram again.

Record Write what you know about the recycling bin.

The top part of the bin is a _____ prism.

The top part is _____ feet long, _____ feet high, and _____ feet wide.

The bottom part of the bin is a _____ prism.

The bottom part is _____ feet long and has a triangular face that is a right triangle, with legs _____ feet and _____ feet.

Think about how to use what you know to solve this problem.

Decide What to Do

You know that the bin is made up of two different shapes. You know the lengths of all the sides of the shapes.

Ask How can I find the volume of the bin?

- I can use the strategy *Solve a Simpler Problem*.

- I need to find the volumes of the rectangular prism and the triangular prism. When I add these volumes, I can find the volume of the bin.

Use Your Ideas

Step 1 Find the volume of the rectangular prism.

$V = lwh$

$= 12$ feet $\times$ _____ feet $\times$ _____ feet

$=$ _____ cubic feet

To visualize this problem, it may help you to draw and label both prisms.

Step 2 Find the volume of the triangular prism.

$V =$ area of the triangular base $\times$ height of the prism

$V = \frac{1}{2} \times$ (base $\times$ height) $\times$ 12 ft

$= \frac{1}{2} \times$ _____ ft $\times$ _____ ft $\times$ 12 ft

$=$ _____ cubic feet

Step 3 Add the volumes to find the volume of the bin.

Volume of Rectangular Prism	+	Volume of Triangular Prism	=	Volume of Bin
_____	+	_____	=	_____

The volume of the bin is _____ .

Review Your Work

Check that you correctly labeled the units in your answer.

Describe How did solving a simpler problem help you?

_____

_____

Try It

Solve the problem.

(1) A town builds a compost bin that is a rectangular prism.
It measures 27 feet long, 9 feet wide, and 3 feet high. The bin
is filled to the top with yard waste in the fall. Composting
can reduce yard waste volume by about $\frac{3}{4}$. About how much
finished compost does the town have in the spring?

Mark
the Text

☐ Read the Problem and Search for Information ·········

Reread and think about sketching and labeling a diagram.

☐ Decide What to Do and Use Your Ideas ··············

You can use the strategy *Solve a Simpler Problem*. Use compatible
numbers to see what process to follow.

Step 1 Write the formula for volume of a rectangular prism.
Substitute the dimensions of the bin for l, w, and h.

Step 2 Simplify the problem. Use a number for the length that is
divisible by 4 to make a simpler problem.
Find $\frac{1}{4}$ of $l \times w \times h$, which is $\frac{1}{4} \times 28 \times 9 \times 3$.

The volume of the town's finished compost is about
_____ cubic feet.

Ask
Yourself

What fraction of the
original volume is
the final volume?

Step 3 Solve the problem with the original numbers. Round your
answer to the nearest cubic foot. Find the product
$\frac{1}{4} \times$ _____ $\times$ _____ $\times$ _____ .

The volume of the town's finished compost is about
_____ cubic feet.

☐ Review Your Work ·····························

Check your multiplication in Step 3.

Clarify How did using compatible numbers help you?

_____

_____

Apply Your Skills

Solve the problems.

② Janine buys a compost bin to put in her backyard. It has the shape shown. What is the volume of the compost bin?

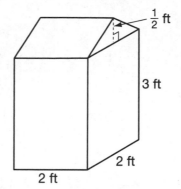

½ ft

3 ft

2 ft

2 ft

◄ **Hint** Think of the bin as having two parts.

Ask Yourself

What shapes is the bin made of?

Volume of Rectangular
Prism = 2 × 2 × _____

Volume of Triangular
Prism = $\frac{1}{2}$ × _____ × _____ × 2

Answer _____

Determine Suppose the bin is filled to half the height. Could you divide the total volume by 2 to find out how much compost is in the bin? Explain.

_____

_____

Ask Yourself

What equation can I write to make this problem simpler?

③ Containers and packaging are a little less than $\frac{1}{3}$ of the waste produced by a certain building. The trash bin at this building is a rectangular prism. It is 70 inches high, 58 inches wide, and 71 inches long. About what volume of the trash in a full bin is containers and packaging?

Volume = length × width × height

Volume of full bin = _____ × _____ × _____

Volume of containers and packaging = ___ × ___ × ___ × ___

◄ **Hint** Use the formula for finding the volume of a rectangular prism.

Answer _____

Retell What can you do to make the problem simpler?

_____

_____

4 Jamie collects shredded paper in the recycling bin shown. So far, she has filled the bin with 2 cubic feet of shredded paper. How much more shredded paper can the recycling bin hold?

Hint Break the shape into a square prism and two congruent triangular prisms.

Ask Yourself

What is the length of the base of the triangle?

2 ft

2 ft

2 ft

2 ft

1 ft

1 ft

Volume of square prism = _____ × _____ × _____

Volume of one triangular prism = $\frac{1}{2} \times$ _____ $\times$ _____ $\times$ _____

Answer _____

Examine A student says that the bin can hold 3 more cubic feet of shredded paper. What mistake might the student have made?

_____

_____

5 Suppose one ton of paper takes up 90 cubic feet of space. The transfer station has a large recycling bin where people take their recycling. It is a rectangular prism 21 feet long, 6 feet wide, and 6 feet high. About how many tons of paper can it hold?

Hint One cubic foot measures 1 ft × 1 ft × 1 ft.

Ask Yourself

How can I make this problem simpler?

Volume = _____ × _____ × _____

Answer _____

Analyze How did you know which operation to use?

_____

_____

On Your Own

Solve the problems. Show your work.

6 Compostable waste makes up about $\frac{1}{4}$ of the total waste in Greenville. The total waste fills a trailer every day. The trailer is a rectangular prism 7 feet wide, 9 feet high, and 21 feet long. What is the volume of the non-compostable waste in Greenville every day?

Answer _____

Support How did you know what to do with the volume of the trailer?

_____

_____

7 The recycling center has a large bin, as shown. The center collects millions of glass bottles each year and crushes them. Philip estimates that about 200 glass bottles can fit in 1 cubic meter. About how many glass bottles can fit in the bin?

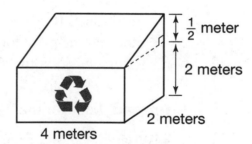

$\frac{1}{2}$ meter

2 meters

2 meters

4 meters

Answer _____

Analyze What information is given that is *not* needed to solve the problem?

_____

_____

Create Write a problem about the volume of a bin at a recycling center that can be solved using the *Solve a Simpler Problem* strategy. Solve your problem.

Strategy Focus
Draw a Diagram

MATH FOCUS: Transformations and Symmetry

Learn About It

▣ Read the Problem ....

Wanda's art class is studying block prints. The block must be a reflection of what each student wants his or her print to look like. A sample print and block is shown below.

Sample

S H O P   ꟼ O H Ƨ

Print   Block

Wanda will make a block print sign for a school bike sale fundraiser. The sign will say "USED BIKES." She will design the block so the print looks like the block letters above. Which of the letters will look the same in the print as on the block?

Reread Look for the important information.

● What is Wanda's project?

_____

● What question will you answer when you solve the problem?

_____

Mark the Text ✏

▣ Search for Information ....

Read the problem again. Then mark or find the information you need for each part of the problem.

Record Write the information that will help you solve the problem.

The block must be a _____ of the print.

In the sample block, the letters _____ and _____ look the same as in the print.

This information will help you answer the question.

Decide What to Do

You know the text of Wanda's print. You know you need to reflect that text to make the block. You also know that you can answer the question after you design the block.

Ask How can I find which letters will look the same in the print?

● I can use the strategy *Draw a Diagram*.

● Then I can compare the letters in the image and its reflection to see which ones look the same.

Use Your Ideas

Step 1 Draw a vertical line of reflection. You will reflect the image over this line. Draw a diagram of your block on the grid below.

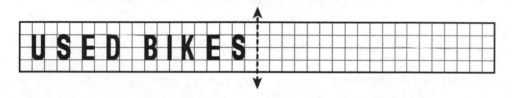

Be careful to reflect the letters correctly over the vertical line.

Step 2 Study the letters in the print and in your diagram of the block to decide how to answer the question.

The letters _____ look the same in the print and the block.

Review Your Work

Check to be sure you did not miss any matching letters.

Describe How does drawing a diagram help you solve this problem?

_____

_____

Try It

Solve the problem.

(1) Kevin is designing a mural for a wall. He draws Triangle 1 as shown. He reflects the triangle over the vertical dashed line, then over the horizontal dashed line, and finally again over the vertical dashed line. After each reflection, he draws the new triangles and labels them Triangle 2, 3, and 4, in that order. Describe the symmetries in the design.

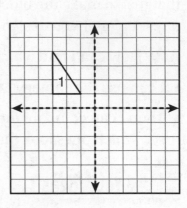

Mark the Text --->

Read the Problem and Search for Information · · · · · · · · ·

Study the diagram to see how the information can help you.

Decide What to Do and Use Your Ideas · · · · · · · · · · · ·

You can use the strategy *Draw a Diagram*.

Step 1 Draw the result of reflecting Triangle 1 over the vertical dashed line. Label it Triangle 2. Draw the result of reflecting Triangle 2 over the horizontal dashed line. Label it Triangle 3. Draw the result of reflecting Triangle 3 over the vertical dashed line. Label it Triangle 4.

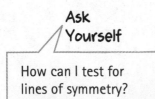

How can I test for lines of symmetry?

Step 2 Look at the completed design to see if it has lines of symmetry or rotational symmetry.

The design has _____ lines of symmetry.

The design has rotational symmetry of _____ .

Review Your Work ·

Imagine folding the completed design along the lines of symmetry.

Clarify How would tracing the final drawing help you?

_____

_____

Apply Your Skills

Solve the problems.

2 Your job is to put up the welcome message for the next softball game. Since this is done from behind the signboard, you must put up the letters backward. Your message is WE ARE NUMBER ONE. Which letters will you need to flip before you put them up?

Reflect each letter over a vertical line.

Letters I need to flip:

_____

Letters I do not need to flip:

_____

Answer _____

Interpret   Why do you *not* have to flip all of the letters?

_____

_____

Ask Yourself

I know this is a reflection. How does that help me solve the problem?

◄ **Hint** A diagram can help you think about these flipped letters.

3 The floor of a community bandstand is a regular octagon divided by 4 diagonals into 8 equal sections. Four of the sections are white and 4 are grey, as shown. How many lines of symmetry does the bandstand floor have? Does the bandstand floor have rotational symmetry? If it does, how many degrees do you have to turn it before it looks the same? If not, explain.

When you fold along a _____ , the two parts match exactly, including the colors.

Ask Yourself

How can I test for line symmetry?

◄ **Hint** To test for rotational symmetry, trace the diagram of the floor and shade the grey areas.

Answer _____

Recognize   Why doesn't the bandstand floor have 4 lines of symmetry?

_____

_____

(4) A group of students is using letter stamps. Each stamp shows a lowercase letter reflected across a vertical line. There is one stamp for each letter. Which stamps will look the same as the letters that they make?

Hint Be careful to write each letter clearly, so that you can compare it to its reflection.

Write out the alphabet in lowercase letters.

_____

_____

Ask Yourself

Which letters look the same when I reflect them?

Answer _____

Explain Why is it important to know that the letters are lowercase?

_____

_____

Ask Yourself

How can I determine if there will be no gaps?

(5) An artist is making a design on tiles that will fill a wall at the town hall. Each tile is the same size. Each tile has the shape of a regular hexagon. Can the artist cover the wall with no gaps between the tiles? Explain.

Each angle of a regular hexagon is _____ degrees.

Hint Draw a diagram. Translate, rotate, or reflect the tiles as needed to cover a wall.

Answer _____

Conclude What types of symmetry does each tile have?

_____

_____

On Your Own

Solve the problems. Show your work.

6 Here is part of the plan for a garden in the park. The garden is centered on a square fountain with walkways leading up to it. The plan has a horizontal and a vertical line of symmetry. Will the garden have rotational symmetry? If it does, how many degrees can you turn it before it looks the same? If not, explain.

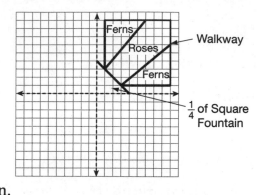

Ferns

Roses

Walkway

Ferns

$\frac{1}{4}$ of Square Fountain

Answer _____

Analyze How could you check the plan for rotational symmetry?

_____

_____

7 A community group is making a new design that includes their motto. To make the design they hold a mirror up to the page below the line of text, TOUGH AS THE HIDE OF AN OX. Which letters will look the same as in the original line of text?

Answer _____

Contrast How is drawing the reflection for this problem different from others in this lesson?

_____

_____

Create

Write a problem about a design with symmetry for which the strategy *Draw a Diagram* could be used. Solve your problem.

In this unit, you worked with four problem-solving strategies. You can use many different strategies to solve a single problem. So if a strategy does not seem to be working, try a different one.

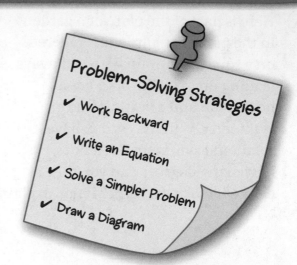

Problem-Solving Strategies

✔ Work Backward

✔ Write an Equation

✔ Solve a Simpler Problem

✔ Draw a Diagram

Solve each problem. Show your work. Record the strategy you use.

1. A checkerboard has 64 dark and light squares. They are arranged in 8 rows of 8. No two squares of the same color are next to each other. The lower left square is dark. What colors are the squares in the other corners?

2. Jimmy is following a treasure map. He starts at an oak tree and walks 10 yards north. Then he turns right and walks 15 yards perpendicular to his first path. Finally, he walks 10 yards south, parallel to his first path. How far is Jimmy from the oak tree?

Answer _____

Strategy _____

Answer _____

Strategy _____

3. A sign in the park says that the walking path is 1,720 feet long. The path is a rectangle. The length of the rectangle is 400 feet. What is its width?

Answer _____

Strategy _____

4. The area of the flat bottom of a rectangular swimming pool at the park is 288 square feet. It is 3.5 feet deep. What is its volume?

Answer _____

Strategy _____

5. You are installing a heater in a greenhouse. The greenhouse has the shape shown. You need to know its volume to find a heater of the right size. What is the volume of the greenhouse?

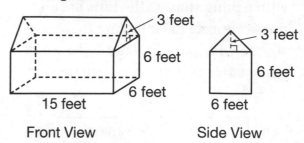

Front View   Side View

Answer _____

Strategy _____

Explain how you knew the length and width of the triangular prism.

Solve each problem. Show your work. Record the strategy you use.

6. The diagram shows a top view of a recycling center. The recycling center wants to cover the entire area between its three collection bins with paving stones. The bins are all the same size. Each paving stone is a square with sides that are 1 foot long. How many paving stones are needed?

```
      ←—12 ft —→|← 4 ft →|
  4 ft ┌─────────┐  ┌─────────┐
       │  Paper  │  │ Plastic │
       └─────────┘  └─────────┘ ┬
                                │ 12 ft
           ┌─────────┐          │
           │  Metal  │          ┴
           └─────────┘
```

Answer _____

Strategy _____

7. You are tiling a floor in this pattern. The tiles are squares in two different sizes. Each large square has an area of 196 square inches. What is the side length of a large square? Of a small square?

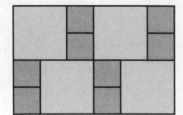

Answer _____

Strategy _____

8. Ambulances use mirror writing so drivers can read the words the right way in their rearview mirrors. The words are written backward and the letters are reversed. Sandy is using mirror writing to paint the word EMERGENCY. Which letters will not look reversed?

Answer _____

Strategy _____

Explain why some letters do *not* look reversed when written in mirror writing.

9. A circular fountain has a circumference of 62.8 feet. Abby is stringing lights straight across the center. About how long is each string of lights? Use 3.14 for π.

Answer _____

Strategy _____

10. Marcy has a new toy box. It has the shape of a rectangular prism. Its dimensions are 35.2 centimeters by 22 centimeters by 60 centimeters. What is the volume of the toy box?

Answer _____

Strategy _____

Write About It

Look back at problem 6. Describe how you used the information in the diagram to help you choose a strategy for solving the problem.

_____

_____

_____

Work Together: Design a Quilt

Your team is making four quilts to hang in the library. You want to use two different colors of congruent isosceles right triangles for each quilt. Work with your team to design four quilts using translations, rotations, and reflections of your triangles. Use 32 triangles to make each quilt.

Plan
1. Draw your isosceles right triangle.
2. Make copies of your triangle using two different colors. Arrange them in different ways using translations, rotations, and reflections. Use 32 triangles for each arrangement.

Decide   As a group, choose four arrangement to make four quilts.

Create   Make a colored diagram of each quilt.

Present   As a group, share your diagrams. Explain how you used translations, rotations, and reflections.

UNIT 5

Problem Solving Using Data, Graphing, and Probability

Unit Theme:
Nature

The world is full of nature. From tiny bugs to huge elephants, living things are everywhere. You may spot reptiles crawling through the grass. You might notice birds flying high in the sky. In this unit, you will see how math is a part of nature.

Math to Know

In this unit, you will use these math skills:

- Make and use bar graphs, line graphs, and circle graphs
- Find the mean, median, mode, and range for a set of data
- Apply probability concepts to dependent and independent events

Problem-Solving Strategies

- Make a Graph
- Make an Organized List
- Work Backward

Link to the Theme

Write another paragraph about Jeremy's poll results. Include some of the facts from the table at the right.
Jeremy loves reptiles! He wants to know what kinds of reptiles his classmates like the most. He lists the top three results in a table.

Reptile	Number of Students
Crocodiles	7
Lizards	5
Snakes	12

_____

_____

_____

_____

_____

_____

Use Math Language

Review Vocabulary

The list below shows vocabulary terms in this unit. Knowing the meaning of these terms will help you understand the problems.

compound event independent events median outcome
dependent events mean   mode   range

Vocabulary Activity Word Groups

Math terms are often learned together because they are related. Use terms from the above list to complete the following sentences.

1. The average found by adding numbers and dividing the sum by the number of addends is the _____ .

2. The _____ is the middle number in an ordered list of numbers.

3. The number or numbers that appears most often in a set of numbers is called the _____ .

4. The _____ is the difference between the least value and the greatest value in a set of numbers.

Graphic Organizer Word Web

Complete the graphic organizer.

- In the center circle, write the definition of *outcome*.

- In each of the 3 smaller circles, write a vocabulary word that is related to the word *outcome*.

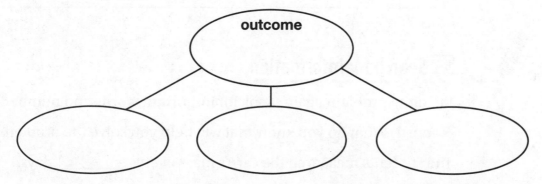

Strategy Focus
Make a Graph

MATH FOCUS: Bar Graphs and Line Graphs

Learn About It

■ Read the Problem

Students in a science class recorded the growth of a carp. The mass of the carp was measured every 20 days. The next four measurements were 300, 360, 420, and 400 grams. In which 20-day period did the carp's mass increase the most?

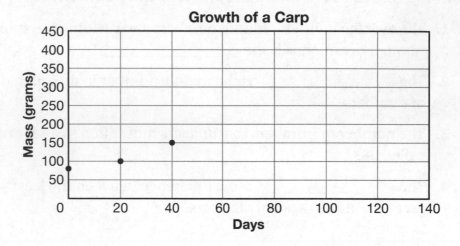

Growth of a Carp

Reread Reread the problem and study the graph.

• What kind of data are provided?

_____

• What is the problem asking me to do?

_____

Mark the Text

■ Search for Information

Read the problem again. Look for important words and numbers.

Record What do you know that will help you solve the problem?

The students measured the carp's mass every _____ days.

The units used to record the mass are _____ .

Think about how you can use the given data to answer the question.

Decide What to Do

You know the change in the mass of the carp over time.

Ask How can I find the 20-day period when the carp's mass increased the most?

- I can use the strategy *Make a Graph* to show all the data.
- I can use the graph to see when the mass increased the most.

A line graph can show change over time.

Use Your Ideas

Step 1 Plot one point for each of the last four measurements given in the problem. Connect all seven points.

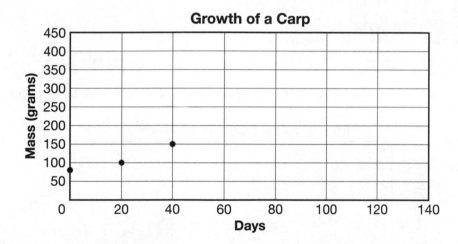

Growth of a Carp

Step 2 Look at the line graph. If the line goes up from left to right, the mass of the carp is _____.

The steepest part of the line shows where the mass of the carp increases the most. The line on the graph is steepest between _____ and _____ days.

So the carp's mass increased the most in the 20-day period between days _____ and _____ .

Review Your Work

Check that each point on the graph is plotted correctly.

Describe How does making a graph help you answer the question?

_____

_____

Try It

Solve the problem.

(1) The bar graph below shows how much time some different kinds of insects spend as eggs. After hatching, honeybees and fruit flies each spend 7 days as larva, silk moths spend 23 days as larva, and lice spend 8 days. Which kind of insect has the greatest difference between time spent as an egg and time spent as a larva? What is that difference?

Mark the Text

▦ Read the Problem and Search for Information ....

Reread the problem carefully.

▦ Decide What to Do and Use Your Ideas ....

Use the strategy *Make a Graph* to show and compare the data.

Step 1   Use the data in the problem to make a double bar graph.

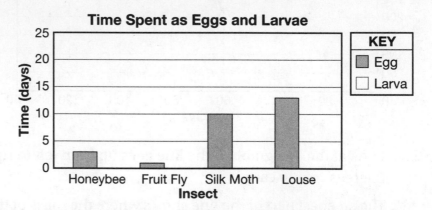

Time Spent as Eggs and Larvae

Ask Yourself

What operation do I use to find the difference?

Step 2   Compare the values on the double bar graph.

The greatest difference between the two bars for one insect is for the _____ .

The difference between the two bars is _____ days.

▦ Review Your Work ....

Check that the number of days is shown correctly for each insect.

Tell How does making a graph help you compare the data?

_____

_____

Apply Your Skills

Solve the problems.

Ask Yourself

Why is it better to display this data in a bar graph rather than a line graph?

② The graph below shows some types of animals that are listed as threatened in the United States. Ten insects, 24 reptiles, and 11 snails are also listed as threatened. There are 3 times as many reptiles listed as threatened as which other animal?

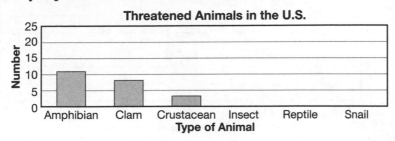

Threatened Animals in the U.S.

◀ **Hint** You can compare the bars to answer the question.

The bars for _____ and _____ are the same height.

Answer _____

Explain How did you decide on the height to shade each bar?

_____

_____

③ Dave and Kyra measure rainfall in Chicago and in Houston. In November Dave measured 2.9 inches. In December he measured 2.8 inches. Kyra measured 4 and 3.7 inches in the same months. In which months did it rain more in Chicago?

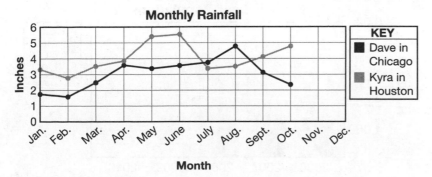

Monthly Rainfall

KEY
■ Dave in Chicago
■ Kyra in Houston

◀ **Hint** A double line graph will work best for this problem.

Ask Yourself

How can I use the graph to help me find the answer?

In November and December, it rained more in _____ .

Answer _____

Determine Why is a line graph a good choice? Explain.

_____

_____

④ The graph shows some species that are endangered. There are also 14 moth and 3 beetle species that are endangered. Which three insects combined have the same number of endangered species as one other kind of insect?

Ask Yourself

Do I need to display the data in any special order?

Hint Think about combining the heights of three bars.

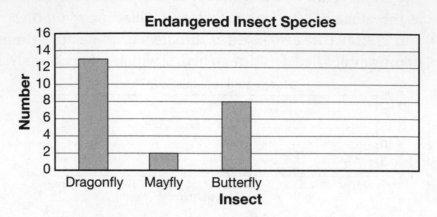

Answer _____

Decide How did you choose which three insects to add together?

_____

_____

⑤ The graph shows estimates for the numbers of elk and deer in a national park in 2006, 2007, and 2008. In 2009, there were an estimated 6,000 elk and 15,800 deer. In 2010, there were an estimated 12,100 elk and 18,700 deer. In which years was the number of elk more than half the number of deer?

Ask Yourself

How can I use the graph to find the answer?

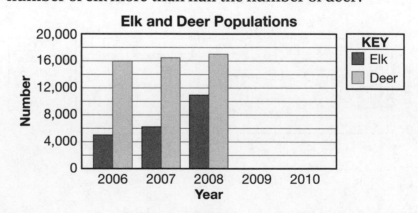

Hint Be sure to compare the populations in each year.

Answer _____

Examine Have you seen a problem like this before in this lesson? Explain.

_____

_____

On Your Own

Solve the problems. Show your work.

(6) Students counted the number of animals they saw at a nearby pond. They counted 13 frogs, 20 ducks, 16 turtles, 13 minnows, 21 dragonflies, and 30 water striders. Which animal count is about twice as many as another?

Answer _____

Formulate What is another question you can ask about the animal count?

_____

_____

(7) Scientists sort and weigh acorns that fall from each of two oak trees. A red oak dropped 78, 92, 80, 70, 80, and 66 ounces of acorns over 6 years. A black oak dropped 50, 90, 80, 70, 88, and 75 ounces of acorns over the same 6 years. In which year was there the greatest difference between the number of acorns dropped by the two trees?

Answer _____

Judge How is this problem like Problem 1?

_____

_____

Create

Use the data from Problem 1 to write a different problem that can be solved using the strategy *Make a Graph*. Solve your problem.

Strategy Focus
Make an Organized List

MATH FOCUS: Mean, Median, Mode, and Range

Learn About It

☐ Read the Problem ....

> José and his class visit a wildlife park. They get information about the heights of some of the animals they see. The heights are measured to the shoulders of the animals. The cheetah's height is 2 feet. The giraffe's height is 18 feet. The leopard's height is 3 feet. The zebra's height is 5 feet. The warthog's height is 2 feet. When asked to describe the typical shoulder height for this group of animals, José says 2 feet. Mario says 3 feet. Jenny says 6 feet. What measures did they use to describe the data?

Reread Think about the problem situation. Ask yourself questions as you read.

● What information is provided about the animals?

_____

● What question did the students answer?

_____

● What do you have to find?

_____

Mark the Text ✏--→

☐ Search for Information ....

Read the problem again. Circle the important information.

Record List the information you need.

What are the shoulder heights of the animals?

Cheetah: _____   Giraffe: _____

Leopard: _____   Zebra: _____

Warthog: _____

What numbers did each student use to describe a typical height?

José: _____ feet   Mario: _____ feet   Jenny: _____ feet

Think of how you can organize the data to solve the problem.

▨ Decide What to Do

You know the shoulder heights of the animals. You also know how the students described the typical height.

Ask How can I find which of the measures each student used?

- I can use the strategy *Make an Organized List*.

- I can use my list to find the mean, median, and mode.

The mean, median, and mode are measures that are often used to describe data.

▨ Use Your Ideas

Step 1 List the shoulder heights of the animals. Then order them from least to greatest.

2, 18, 3, 5, 2

2, _____ , _____ , _____ , 18

Step 2 Find the **mean**.

2 + 2 + 3 + 5 + 18 = _____

_____ ÷ 5 = _____

Step 3 Find the **median**. Circle it.

2, 2, 3, 5, 18

Step 4 Find the **mode**.

What height occurs most often in your list? _____

José said a typical shoulder height was 2 feet. He described the data using the _____ .

Mario said a typical shoulder height was 3 feet. He described the data using the _____ .

Jenny said a typical shoulder height was 6 feet. She described the data using the _____ .

▨ Review Your Work

Check that your list includes all of the heights.

Identify Which two measures was your organized list most helpful in finding?

_____

_____

Try It

Solve the problem.

	Birth Weight (pounds)	Adult Weight (pounds)
Hippo A	8	450
Hippo B	9	375
Hippo C	13	550
Hippo D	11	425
Hippo E	10	380
Hippo F	12	520
Hippo G	9	490

① Vicky studies the growth of pygmy hippos. The birth weights and adult weights are shown in the table. Which hippo had the median birth weight? Which had the median adult weight?

Mark the Text

▢ Read the Problem and Search for Information ┊ . . . .

Restate the problem in your own words. Make sure you understand the information given in the table.

▢ Decide What to Do and Use Your Ideas ┊ . . . .

You can *Make an Organized List* to find the medians.

Ask Yourself

How must I organize the numbers to find the median?

Step 1 Find the median of the birth weights. Write the weights in order from least to greatest. Circle the median.

8, _____ ,13

Step 2 Find the median of the adult weights. Write the weights in order from least to greatest. Circle the median.

375, _____ , 550

So the median birth weight belongs to Hippo _____ and the median adult weight belongs to Hippo _____ .

▢ Review Your Work ┊ . . . .

Check that you recorded the correct number of weights.

Predict Dean arranged the weights from greatest to least. Will he get the correct answer? Explain.

_____

_____

Apply Your Skills

Solve the problems.

(2) The African Savanna has a wet climate in the summer and a dry climate in the winter. Barry recorded the average high temperatures in the Savanna for 12 months. The temperatures were 70°F, 74°F, 72°F, 76°F, 81°F, 82°F, 85°F, 84°F, 78°F, 82°F, 76°F, and 76°F. What is the median of the data?

70, _____ , 85

Median: $(76 +$ _____ $) \div 2 =$ _____

Answer _____

Locate What given information is *not* needed?

_____

_____

◀ **Hint** First write the numbers in order from least to greatest.

Ask Yourself

How do I find the median when there is an even number of numbers?

(3) The number of people signed up for each of Saturday and Sunday's five safari tours are shown below. Which day had the greater median number of visitors? The greater range?

Tour	Visitors on Saturday	Visitors on Sunday
10 A.M.	23	26
11 A.M.	40	38
12 Noon	28	37
1 P.M.	24	40
2 P.M.	35	31

◀ **Hint** The times of the tours are not important for this problem.

Ask Yourself

Do I need to use data for both days, or just one?

List the numbers in order. Circle the medians and find the ranges.

Saturday: 23, _____ , 40   Range: 40 − 23 = _____

Sunday: 26, _____ , 40   Range: 40 − 26 = _____

Answer _____

Relate What is another question you could ask about the data?

_____

_____

4 Dr. Lu is studying prides of lions. She counts the numbers of male and female lions in six different prides. How do the mean, median, and range of the numbers of females compare to the numbers of males in the prides?

Males	4	3	3	2	4	2
Females	8	17	10	5	12	14

List the numbers in order from least to greatest.

Male Lions: _____

Female Lions: _____

	Mean	Median	Range
Males			
Females			

Answer _____

Determine Suppose the question only asked how much greater the mean number of female lions was than the mean number of male lions. Would you have made an organized list to answer the question? Explain.

_____

_____

5 Maggie studies African wild dog populations. She records the number of pups in several litters. The numbers are 7, 8, 17, 10, 14, 12, 20, 14, 8, and 10. Maggie says she could use the numbers 8, 10, 11, 12, or 14 to describe a typical litter size. What measures do these numbers represent?

List the numbers in order from least to greatest.

Mean	Median	Modes

_____

Answer _____

Explain How did making an organized list help you solve the problem?

_____

_____

On Your Own

Solve the problems. Show your work.

6 A team of biologists is studying baby elephants. They record the weights of 10 babies. Their weights are 230, 250, 210, 210, 240, 220, 225, 200, 210, and 205 pounds. When asked to describe the typical weight Dr. Ryan says 210 pounds. Dr. Suarez says 215 pounds. Dr. Kane says that a typical weight is 220 pounds. Was Dr. Kane using the mean, median, or mode?

Answer _____

Analyze Kelly thinks that the median of the data is 230. What mistake might Kelly have made?

_____

7 Brian is writing a report on chimpanzees. He finds data about weights of adult chimpanzees. What mean, median, and range should Brian use in his report for the weights of male and female chimps?

Male Weights (pounds)	Female Weights (pounds)
90	70
120	95
100	100
95	80
110	90

Answer _____

Plan How would you organize the data differently if you needed to find the mean, median, and range of all of the chimpanzees?

_____

_____

Create Survey your classmates to find the numbers of different types of pets they have. Write a problem that can be solved using the strategy *Make an Organized List*. Solve your problem.

Strategy Focus
Work Backward

MATH FOCUS: Circle Graphs

Learn About It

▢ Read the Problem

Mia's class is learning about reptiles. They counted reptiles at a pond next to their school. Many turtles and lizards live there. The circle graph shows the reptiles they counted. How many lizards did Mia's class count?

Reptiles at the Pond

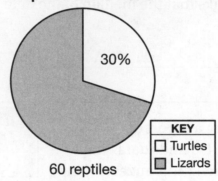

30%

60 reptiles

KEY
☐ Turtles
▨ Lizards

Reread Keep these questions in mind.

• What is the problem about?

_____

• What does the circle graph show?

_____

• What do I need to find?

_____

Mark the Text

▢ Search for Information

Look at the problem and the circle graph. Mark the different information you need.

Record Write the information you need to solve the problem.

_____ % of reptiles the class counted are turtles.

The class counted a total of _____ reptiles.

Think about how you can use this information.

Decide What to Do

The circle graph shows only part of the information. You need to find more data to answer the question.

Ask How can I find the number of lizards Mia's class counted?

- I know how many reptiles they counted. I know the percent of those reptiles that are turtles.

- I can use the strategy *Work Backward* to find the number of lizards they counted.

Use Your Ideas

Step 1 Find the number of turtles they counted.

$$30\% \text{ of } 60 = 0.30 \times 60 = \underline{\hspace{2cm}}.$$

Reptiles at the Pond

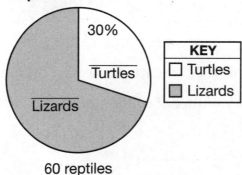

30%

Turtles

Lizards

KEY
- ☐ Turtles
- ▨ Lizards

60 reptiles

As you find numbers that help you solve the problem, write them in the circle graph.

Step 2 Use the number of turtles to find the number of lizards.

$$60 - \underline{\hspace{2cm}} = \underline{\hspace{2cm}}$$

So Mia's class counted _____ lizards.

Review Your Work

Check that the number of lizards and the number of turtles the class counted add up to a total of 60 reptiles.

Describe How did you know how to find the number of turtles the class counted?

_____

_____

Try It

Solve the problem.

Crocodile Eggs

① Park rangers are counting crocodile eggs. Crocodile eggs were found in the two areas shown in the graph. What percent of the eggs were found in the East Area of the park?

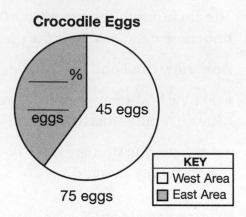

_____%

_____ eggs

45 eggs

75 eggs

KEY
☐ West Area
▨ East Area

Mark the Text

▨ Read the Problem and Search for Information ....

Think about the information you need to find.

▨ Decide What to Do and Use Your Ideas ....

A total of _____ eggs were found in the park.

Use the strategy *Work Backward* to solve the problem.

Step 1 Find the number of eggs found in the East Area of the park.

75 − 45 = _____

Step 2 Find the percent of eggs found in the East Area of the park.

_____ is _____ % of 75.

So _____ % of the eggs were found in the East Area of the park.

Ask Yourself

How can I find the percent of the eggs that were found in the East Area of the park?

▨ Review Your Work ....

Check that the total number of eggs found in the park is 75.

Explain Why is *Work Backward* a good strategy to use?

_____

_____

Apply Your Skills

Solve the problems.

(2) A local group counts animals in a park. The circle graph shows the population of female and male box turtles in the park. What percent of the box turtles are female?

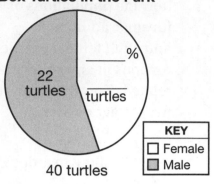

Box Turtles in the Park

_____%

22 turtles

_____ turtles

40 turtles

KEY
- ☐ Female
- ■ Male

Ask Yourself

How do I find the missing percent?

Find the number of female box turtles. 40 − 22 = _____

Find the percent that are female. _____ out of 40 is _____ %.

Answer _____

Compare How could you use percents to check your answer?

_____

_____

Hint To find the percent one number is of another, divide the first number by the second number and change the result to a percent.

(3) The circle graph shows the relationship between the number of alligators and crocodiles in an exhibit. How many more crocodiles than alligators are on display?

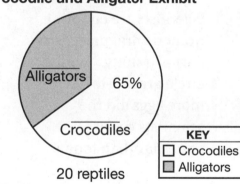

Crocodile and Alligator Exhibit

_____ Alligators

65%

_____ Crocodiles

20 reptiles

KEY
- ☐ Crocodiles
- ■ Alligators

Find the number of crocodiles. 65% of 20 is _____ crocodiles.

Then find the number of alligators.
20 − _____ = _____ alligators.

Answer _____

Determine What words tell you what operation to use? Explain.

_____

_____

Ask Yourself

How do I find the percent of a number?

Hint Be sure to answer the question that is asked by the problem.

(4) The circle graph shows how the populations of juvenile, adult male, and adult female bearded dragons in an exhibit are related. How many adult male dragons there?

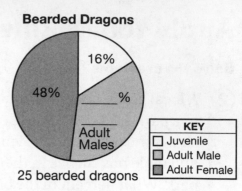

Bearded Dragons

16%

48%

_____ %

Adult Males

25 bearded dragons

KEY
☐ Juvenile
▨ Adult Male
▨ Adult Female

Hint Find the percent of adult male dragons first.

Ask Yourself

How can I estimate a reasonable answer?

_____ % of the bearded dragons are adult males.

Answer _____

Relate How is this problem different from the Learn About It problem in this lesson?

_____

_____

(5) Scientists are counting Komodo dragon eggs as part of a study. Use the circle graph. How many more eggs did the scientists count yesterday than today?

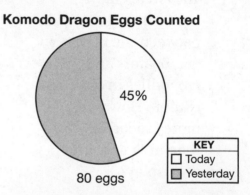

Komodo Dragon Eggs Counted

45%

80 eggs

KEY
☐ Today
▨ Yesterday

Ask Yourself

What percent of the eggs did they count yesterday?

They counted _____ today.
They counted _____ eggs yesterday.

Hint Be sure to answer the question that is asked by the problem.

Answer _____

Analyze How did you find the number of eggs the scientists counted yesterday?

_____

_____

On Your Own

Solve the problems. Show your work.

6 Scientists counted a number of alligators and crocodiles along the banks of a river in Florida. The circle graph shows their results. What percent of the reptiles were alligators?

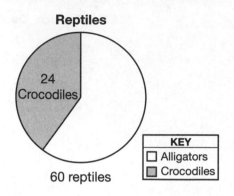

Reptiles

24 Crocodiles

60 reptiles

KEY
☐ Alligators
▨ Crocodiles

Answer _____

Identify How did you find the percent that are alligators?

_____

_____

7 A wildlife refuge is home to both land iguanas and marine iguanas. The circle graph shows how the populations are related. How many more land iguanas than marine iguanas are there?

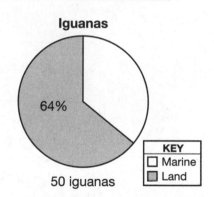

Iguanas

64%

50 iguanas

KEY
☐ Marine
▨ Land

Answer _____

Conclude Describe how working backward helped you solve this problem.

_____

_____

Create Write a problem about poisonous and non-poisonous snakes that can be solved using the *Work Backward* strategy. Solve your problem.

Strategy Focus
Make an Organized List

MATH FOCUS: Dependent and Independent Events

Learn About It

Read the Problem . . . .

Jeremy is playing a game. He spins the spinner shown twice. The spinner is equally likely to land on any of the 5 sections. What is the probability that the spinner lands on an owl both times?

Reread Think of these questions as you read the problem.

• What is the problem about?

_____

• What do you need to find?

_____

Mark the Text

Search for Information . . . .

Read the problem again. Circle the important information in the problem.

Record Write the information you need.

What birds are on the spinner?

_____

Which of the birds are owls?

_____

How many times will Jeremy spin the spinner?

_____

Think about the ways the spinner can land.

Decide What to Do

You know that there are 5 different types of birds on the spinner.
You also know that 3 of them are owls.

Ask What do I need to do?

- I need to find the probability of a compound event. Landing on an owl the first time and landing on an owl the second time are independent events.

- I can *Make an Organized List* to find all of the possible outcomes.

Use Your Ideas

Step 1 Make a list of all of the possible results of the 2 spins.

W = white owl, G = gray owl, T = tan owl, B = bald eagle, H = hawk

(WW) WG, WT, WB, WH

GW, GG, GT, GB, _____

TW, TG, TT, _____ , _____

BW, BG, _____ , _____ , _____

HW, _____ , _____ , _____ , _____

Step 2 Find the favorable outcomes. Circle the choices in Step 1 that have 2 owls.

Step 3 How many of the possible results have 2 owls? _____

How many possible results are there in all? _____

P(2 owls) = ——— ← number of favorable outcomes
← total number of possible outcomes

The probability that the spinner lands on an owl both times is _____ .

Review Your Work

Check that your list includes all possible choices without repeats.

Describe How does making an organized list help?

_____

_____

> Use letters to represent the different birds.

Try It

Solve the problem.

① Noelle has a yellow canary, a red finch, and a yellow finch. She randomly chooses a different cage for each. She has a white birdcage, a black birdcage, and a silver birdcage. What is the probability that she places a yellow bird in the white cage and a finch in the black cage?

Mark the Text

☐ Read the Problem and Search for Information ....

Think about how to use the information to answer the question.

☐ Decide What to Do and Use Your Ideas ....

These are dependent events because after Noelle places one bird, it cannot go in another cage. You can *Make an Organized List*.

Step 1 List all of the ways to place each bird in a cage.

Ask Yourself

What is a good way to list the outcomes?

White Cage	Black Cage	Silver Cage
Yellow Canary	Red Finch	Yellow Finch
Yellow Canary	Yellow Finch	Red Finch
Red Finch	Yellow Canary	
Red Finch		
Yellow Finch		

Step 2 How many arrangements have a yellow bird in the white cage and a finch in the black cage? _____

There are _____ ways to arrange the birds in all.

So the probability that Noelle places a yellow bird in the white cage and a finch in the black cage is _____ .

☐ Review Your Work ....

Check that you have not forgotten any possible arrangements.

Conclude How did a table help you make your list?

Apply Your Skills

Solve the problems.

(2) A museum will choose two different models of extinct birds for an exhibit. One will be placed on the left side of the exhibit. The other will be placed on the right. There are five models to choose from. How many ways can the museum arrange two models for the exhibit?

◀ **Hint** Use the letters A, B, C, D, and E to represent the different models.

A on the left: AB, AC, AD, AE

B on the left: BA, _____ , _____ , _____

C on the left: _____ , _____ , _____ , _____

D on the left: _____ , _____ , _____ , _____

E on the left: _____ , _____ , _____ , _____

Ask Yourself

For this question, is AB the same as BA?

Answer _____

Explain Why is BB not in your list?

_____

_____

(3) A store has 4 bird feeders and 5 bird baths. Each bird feeder is a different color. One is red, one is white, one is brown, and one is green. A customer selects 2 different bird feeders at random. What is the probability that the customer chooses a red bird feeder and a brown bird feeder?

◀ **Hint** The customer cannot choose two of the same color.

Combinations that include red: RW, RB, RG

Combinations that include white, not already listed: WB, _____

Combinations that include brown, not already listed: _____

Combinations that include green, not already listed: _____

Ask Yourself

Is the order in which the bird feeders are chosen important?

Answer _____

Identify What information is given that is *not* needed to solve the problem?

_____

_____

(4) People can sponsor injured birds through a charity program. There are 5 wild turkeys waiting for sponsors. The 3 males are white, gray, and brown. The 2 females are white and brown. If the birds are chosen at random, what is the probability that the first turkey sponsored is male and the second is white?

Hint Use two letters to represent each turkey, such as "WM" for "white male."

Ask Yourself

If a white male is chosen first, which birds are left to be chosen second?

WM first: _____

GM first: _____

BM first: _____

WF first: _____

BF first: _____

Answer _____

Analyze Why did you need to write combinations where a female was first?

_____

_____

(5) Bonnie chooses T-shirts to sell for a fundraiser. The shirts come in blue, white, and black. The T-shirt will have a picture of a cardinal, an oriole, or a blue jay. If Bonnie chooses a color and picture combination for the shirts at random, what is the probability she will choose a shirt that is blue, a shirt that has a picture of a blue jay, or a shirt that is blue and has a picture of a blue jay?

Color	Bird

Hint List the outcomes in a table.

Ask Yourself

Can a shirt be blue *and* have a picture of a blue jay?

Answer _____

Examine Yuri thinks that the probability of choosing a blue shirt or a shirt with a blue jay is $\frac{2}{3}$. What mistake might Yuri have made?

_____

_____

On Your Own

Solve the problems. Show your work.

6 There are five trading cards on a table. Each card shows a picture of a different bird. The birds shown are emu, pelican, robin, hummingbird, and finch. Bill chooses 2 cards at the same time without looking. What is the probability that he chooses the emu and the finch?

Answer _____

Interpret Suppose Bill chooses the cards one at a time. Can you use the same list to find the probability that Bill chooses the emu first and the finch second? Explain.

_____

_____

7 A bird shelter is celebrating National Bird Day. As part of the festivities, a ranger will show 6 different birds, one at a time. There will be an owl, a parrot, a flamingo, a quail, an egret, and a heron. What is the probability that the ranger will first show the owl and then the parrot?

Answer _____

Relate How would your list be different if you need to find the probability of showing the birds in the order owl, parrot, flamingo, quail, egret, and heron?

_____

_____

Create

On a bird watch, you see 6 birds in a pond. There are 3 egrets, 2 geese, and 1 duck. Write your own problem that can be solved using the strategy *Make an Organized List*. Solve your problem.

Review What You Learned

In this unit, you worked with three problem-solving strategies. You can often use many different strategies to solve a single problem. So if a strategy does not seem to be working, try a different one.

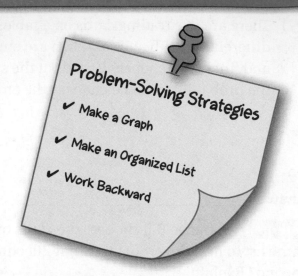

Problem-Solving Strategies

✔ Make a Graph

✔ Make an Organized List

✔ Work Backward

Solve each problem. Show your work. Record the strategy you use.

1. Last spring, Ari and four friends helped the local conservation officer count salamanders. They started to make a bar graph to show how many salamanders they counted. Larry counted 12 salamanders. Terry counted 14. Who counted about half as many salamanders as Harry?

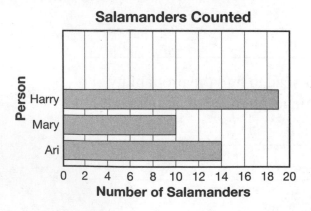

Salamanders Counted

2. An African safari leader counts the number of elephants she sees for 8 days. What are the mean, median, mode, and range of the data?

Day	1	2	3	4	5	6	7	8
Elephants	13	10	15	30	12	16	13	27

Answer _____

Strategy _____

Answer _____

Strategy _____

3. Jolie's bowling scores were 83, 97, 89, 85, and 96. What was her median score?

Answer _____

Strategy _____

4. Romi kept track of how many robins, sparrows, and cardinals she saw at her birdfeeder. She made a circle graph to show the data. How many robins has she seen?

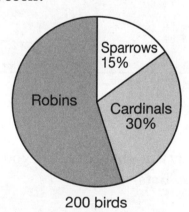

200 birds

Answer _____

Strategy _____

5. Maura collected data about temperatures during one day. Each hour, Maura recorded the temperature on a line graph. At 2 P.M., Maura saw that the temperature was 30° F. The temperature was 29°F at 3 P.M. Predict the approximate temperature reading at 4 P.M.

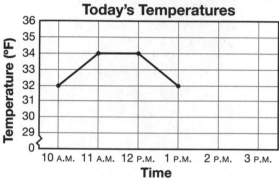

Answer _____

Strategy _____

Explain how you made your prediction.

Solve each problem. Show your work. Record the strategy you use.

6. Bobby makes a circle graph showing the number of pairs of socks he has. What percent of his socks are *not* white, black, or brown?

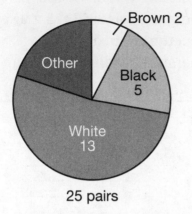

Brown 2

Other

Black 5

White 13

25 pairs

Answer _____

Strategy _____

7. The manager at a local botanical garden kept track of the number of visitors each day for one week. What are the mean, median, and range of the data?

Sun.	Mon.	Tue.	Wed.	Thu.	Fri.	Sat.
850	180	185	235	220	205	750

Answer _____

Strategy _____

8. While traveling, Jaya and Bruce counted colors of different cars. Jaya made a bar graph to show the cars she counted. Bruce counted 75 white cars, 40 blue cars, 25 red cars, and 50 black cars. Which color car shows the greatest difference between the number that Jaya counted and the number that Bruce counted?

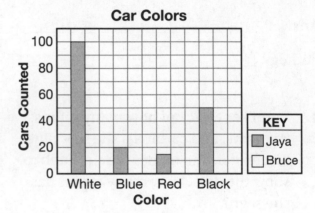

Car Colors

Cars Counted

KEY
Jaya
Bruce

White   Blue   Red   Black
Color

Answer _____

Strategy _____

Explain why a double graph is a good way to show this data.

9. Belle plays a game where she spins 2 spinners. One spinner has 4 equal sections labeled A, B, C, and D. The other spinner has 6 equal sections numbered 1 to 6. What is the probability that Belle spins a B and an even number?

10. Joseph surveyed some students in his class about the numbers of brothers and sisters they have. The results were 1, 3, 2, 1, 2, 1, 1, 3, 4. What are the mean, median, mode, and range of the data?

Answer _____

Strategy _____

Answer _____

Strategy _____

Write About It

Look back at Problem 9. Justify your answer.

_____

_____

_____

Work Together: Choose a Uniform

Your basketball team is choosing uniforms for next season. You have a budget of $1,000. You need jerseys and shorts. You can also buy T-shirts and warm-up pants.

Plan
1. List all the different color combinations of jerseys and shorts. Choose one combination. Choose a color for T-shirts and warm-up pants if your team plans to buy those.

2. Discuss how many of each item your team will buy. Will you buy the same number of each item?

3. Find the total amount your team will spend for each item. Then find the total cost for all of the items.

Create Draw a picture of the uniforms you chose. Add your own school designs. Then make a bar graph showing the amount you will spend on each part of your uniform.

Present As a group, share your uniform with the class. Display your bar graph and explain how you decided how many of each item to buy. Discuss and compare the amounts you will spend on each item.

Basketball Uniforms
Jerseys
$20
Blue, White, Red, Black, Yellow
Shorts
$15
Blue, White, Black
T-Shirts
$10
White, Gray
Warm-Up Pants
$20
Blue, White, Black

Unit Theme:
Discovery

The world is full of things to discover. You can discover things in nearby cities. You can discover things about rainforests that are far away. Scientists make discoveries on dry land and deep under the sea. No matter where you are, there is always something to check out. In this unit, you will see how math is part of discovery.

Math to Know

In this unit you will use these math skills:

- Solve equations and find function rules
- Add and subtract integers
- Plot points on the coordinate plane

Problem-Solving Strategies

- Write an Equation
- Look for a Pattern
- Draw a Diagram
- Make a Graph

Link to the Theme

Write another paragraph about Ling and the Robot Show. Include some of the facts from the table.
Ling wants to buy a ticket for the robot show. It is a 3-day event. She looks at the prices for the show.

Robot Show Prices

1-Day Pass	2-Day Pass	3-Day Pass
$15 per person	$25 per person	$35 per person

_____

_____

_____

_____

_____

_____

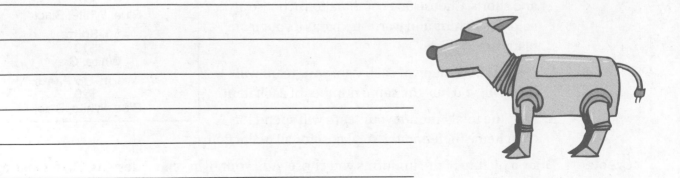

Use Math Language

Review Vocabulary

The list below shows vocabulary terms in this unit. Knowing the meaning of these terms will help you understand the problems.

coordinate grid   function   integer x-axis

equation   function table   inverse operation y-axis

Vocabulary Activity   Modifiers

A descriptive word placed in front of another word indicates a specific meaning. When learning new math vocabulary, pay attention to each word in the term.

1. A _____ matches each input value with one output value.

2. Addition is the _____ of subtraction.

3. A _____ includes an x-axis and a y-axis.

Graphic Organizer   Word Map

Complete the graphic organizer.

- Draw a diagram to show what *equation* means.

- Write a number sentence that is an example of the term.

- Write a number sentence that is *not* an example of an equation.

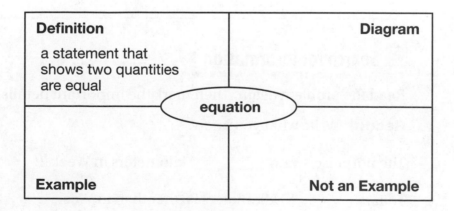

Definition	Diagram
a statement that shows two quantities are equal	
equation	
Example	Not an Example

Strategy Focus
Write an Equation

MATH FOCUS: Equations

Learn About It

■ Read the Problem ....

Dr. Diaz leads a study on porpoises. In Week 1 of the study, she tagged and released a harbor porpoise into the ocean. In Week 2, she started the journal below. Dr. Diaz will record the distances the porpoise swims during the study. How far did the porpoise swim in the first two weeks?

> Week 2   March 15
>
> This week, the porpoise swam a
> distance of 350 kilometers. It
> swam 100 kilometers farther
> this week than in Week 1.

Reread Be sure you understand the main idea of the problem.

● What does Dr. Diaz do?

_____

● What information is given in the problem?

_____

● What do you need to find?

_____

Mark
the Text

■ Search for Information ....

Read the problem again. Then mark the important details.

Record Write what you know.

The porpoise swam _____ kilometers in Week 2.

That is _____ kilometers farther than in Week 1.

Think about how the information is related to the question you must answer.

Decide What to Do

You know the distance the porpoise swam in Week 2. You know how it is related to the distance it swam in Week 1.

Ask How can I find the total distance the porpoise swam in the two weeks?

- I can use the strategy *Write an Equation.*

- I can solve the equation to find the distance it swam in Week 1. Then I can add that distance to the distance it swam in Week 2 to find the total distance.

Use Your Ideas

Step 1 Use d to stand for the number of kilometers the porpoise swam in Week 1. Write an equation to show the relationship between d and the distance it swam in Week 2.

> You can write an equation to represent the problem.

Distance in Week 1	+	Distance farther in Week 2	=	Distance in Week 2
d	+	_____	=	_____

Step 2 Solve the equation.

What number plus 100 equals 350? _____

So the porpoise swam _____ kilometers in Week 1.

Step 3 Find the total distance the porpoise swam.

Distance in Week 1	+	Distance in Week 2	=	Total distance
_____	+	_____	=	_____

So the porpoise swam a total of _____ kilometers.

Review Your Work

Substitute the distance swam in Week 1 for d in the first equation. Check that it works.

Explain Why does the first equation use addition?

_____

_____

Try It

Solve the problem.

(1) Waves can be up to 70 feet tall in some parts of Hawaii. That is the height of a 7-story building! Sam is at a beach in a part of Hawaii where the waves are not that tall. The 70-foot-tall waves are about twice as tall as the waves where Sam is. About how tall are the waves where Sam is?

Mark the Text

▢ Read the Problem and Search for Information ⌐ . . . .

Identify the information you need to find. Reread the problem and circle important numbers.

▢ Decide What to Do and Use Your Ideas ⌐ . . . .

You can use the strategy *Write an Equation* to find the answer to the problem.

Step 1 Use *h* to represent the height, in feet, of the waves where Sam is. Write an equation to show the relationship between *h* and the 70-foot-tall waves.

Ask Yourself

> How are the 70-foot-tall waves related to the waves where Sam is? Should I multiply or divide in my equation?

Height of 70-foot-tall waves	=	Twice the height of the waves where Sam is
_____	=	_____ $\times h$

Step 2 Solve the equation.

What number times 2 equals 70? _____

So the waves where Sam is are about _____ feet high.

▢ Review Your Work ⌐ . . . .

Reread the question. Check that your answer makes sense.

Recognize What information is given that is *not* needed to solve the problem?

_____

_____

Apply Your Skills

Solve the problems.

Ask Yourself

Do I need to add or subtract in my equation?

2 Tara visited a volcano in 2010. Her guide told her that it has been erupting for about 30 years. Tara wants to tell her friend when the volcano started erupting. What year should she tell her friend?

Year eruption began	+	Years erupting	=	Year Tara visited
y	+	_____	=	_____

◀ **Hint** Use y to show the year the volcano started erupting.

Answer _____

Identify What phrase tells you what kind of equation to write?

_____

_____

3 Wild horses live on an island off the coast of North Carolina. They have been there for hundreds of years. At one time, there were about 300 horses on the island. That is about 15 times the number of horses living there now. About how many horses live on the island now?

Ask Yourself

Are there more or fewer horses on the island now?

15	×	Horses now	=	Horses before
_____	×	_____	=	_____

◀ **Hint** Write an equation relating the number of horses on the island now to the number earlier.

Answer _____

Locate What words told you that you needed to multiply in your equation?

_____

_____

Hint The phrase *shared the coins equally* means that everyone received the same number of coins.

(4) Dr. Lee and 3 scientists explored a shipwreck. They found gold coins deep underwater. The scientists shared the coins equally. Each person received 20 gold coins. How many coins did they find in all?

_____ ÷ _____ = _____

Ask Yourself

How many people shared the coins?

Answer _____

Examine Leo thinks that the scientists found 60 coins in all. What mistake could Leo have made?

_____

_____

(5) Ranger Flores is hiking along a cliff on the coast of California. She is 100 feet above sea level when she begins. That is 460 feet less than the highest point on the cliff. The highest point of the cliff is 7 times the height of where she ends her hike. Is Ranger Flores higher when she starts or when she ends? How much higher?

Hint You need to write two equations: one to find the highest point and one to find the ending height.

_____ − _____ = _____

_____ × _____ = _____

Ask Yourself

Can I find the ending height before I find the highest point?

Answer _____

Sequence How did the first equation help you write the second?

_____

_____

On Your Own

Solve the problems. Show your work.

(6) Manny and Rita count the sea lions at 2 different piers in California. Manny counted 400 sea lions. He counted 50 more sea lions than Rita. How many sea lions did they count altogether?

Answer _____

Plan What was the first thing you needed to do to solve the problem?

_____

_____

(7) Ali and her aunt bike along the coast. They start at Gulf Port and stop in Biloxi. Then they bike to Mobile. The distance from Biloxi to Mobile is shown. The distance from Biloxi to Mobile is 5 times the distance from Gulf Port to Biloxi. How many miles longer is their ride from Biloxi to Mobile than from Gulf Port to Biloxi?

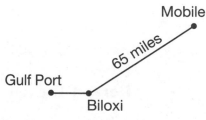

Answer _____

Analyze How did the diagram help you solve the problem?

_____

_____

Create Look at the problems in this lesson. Write a new problem that can be solved by writing an equation. Solve your problem.

Strategy Focus
Look for a Pattern

MATH FOCUS: Functions

Learn About It

🔲 Read the Problem ·

Anna finished building a space satellite kit her brother had started. When she began, 3 pieces were already assembled. The drawings show a pattern for how many pieces she had assembled by the end of 5-minute periods. The pattern continued. Anna put all the pieces of the satellite together in 1 hour. How many pieces is the satellite made of in all?

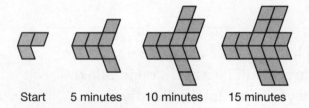

| Start | 5 minutes | 10 minutes | 15 minutes |

Reread Use your own words to tell what the problem is about.

• What is Anna doing?

_____

• What do I need to find?

_____

Mark
the Text

🔲 Search for Information ·

Read the problem again and look at the shapes.

Record What do you know about the pieces in the pattern?

At the start, there were _____ pieces assembled.

After 5 minutes, there were _____ pieces assembled.

After 10 minutes, there were _____ pieces assembled.

After 15 minutes, there were _____ pieces assembled.

Think about how these details help you solve the problem.

Decide What to Do

You know the number of pieces assembled at 5, 10, and 15 minutes. You know that the pattern continues.

Ask How can I find the number of pieces in the satellite?

- I can use the strategy *Look for a Pattern*.

- I can make a function table to show the number of pieces used at different times. Then I can look for a function rule.

Use Your Ideas

Step 1 Make a function table to show the number of minutes and the number of pieces assembled. Identify how the minutes and the number of pieces are related.

Minutes (x)	Number of Pieces (y)
0	3
5	8
10	13
15	
20	

Relationship between x and y

← 5 + _____ = 8

← 10 + _____ = 13

← 15 + _____ = _____

← 20 + _____ = _____

Step 2 The change is the same for each step, so there is a pattern. The function rule is *add* _____ .

Step 3 Write a function rule for the table. $y = x +$ _____

Step 4 Use the rule to solve the problem. You want to find the number of pieces assembled in one hour.

The satellite is made of _____ pieces in all.

There are 60 minutes in 1 hour.

Review Your Work

Check that you correctly described the function.

Explain Why is it easier to solve the problem using the function rule than continuing the pattern in the function table?

_____

_____

Try It

Solve the problem.

1 Mr. Santos is testing the newest electric car. He records how far he drives each time he charges the battery. What is the least number of times will he have to charge the battery to drive at least 800 miles?

Battery Charge	Distance
1	120
2	240
3	360
4	480

Mark the Text

■ Read the Problem and Search for Information ┃ . . . .

Identify the types of information in the problem.

■ Decide What to Do and Use Your Ideas ┃ . . . .

You can use the strategy *Look for a Pattern*.

Step 1 Look at the table. Find how the number of battery charges (*x*) and the distance driven (*y*) are related.

Ask Yourself

What two things can I relate in the function table?

Number of Battery Charges (x)	Distance Driven (y)
1	120
2	240
3	360
4	480

Relationship between x and y

← 1 × _____ = 120

← 2 × _____ = 240

← 3 × _____ = 360

← 4 × _____ = 480

Step 2 Look for a pattern in the function table. Write a rule.

$y = x \times$ _____ Use the rule to solve the problem.

Mr. Santos needs to charge the battery _____ times.

■ Review Your Work ┃ . . . .

Check that you correctly used the rule to complete the table.

Distinguish Why is the strategy *Look for a Pattern* useful?

_____

_____

Apply Your Skills

Solve the problems.

(2) The toy store is having a sale. There are 9 robots at the start of the 12-hour sale. There are 7 robots left with 10 hours to go. There are 5 robots left with 8 hours to go. There are 3 robots left with 6 hours to go. If the pattern continues, how many robots will there be with 3 hours left in the sale?

Hours Left in the Sale (x)	Number of Robots (y)
12	9

Relationship between x and y

12 ⟵   12 $\bigcirc$ _____ = 9

⟵   _____ $\bigcirc$ _____ = _____

⟵   _____ $\bigcirc$ _____ = _____

⟵   _____ $\bigcirc$ _____ = _____

Ask Yourself

What operation can I use to show how the number of robots decreases?

Function Rule: $y = x \bigcirc$ _____

◀ **Hint** You can use a function rule to solve the problem.

Answer _____

Examine How does the table help you to find the function?

_____

_____

(3) Scientists made windmills that fly in the air. A video about these windmills was shown on the Internet. The table shows how many people in all had watched the video by the end of each minute. If the pattern continued, how many people watched the video in the first 7 minutes?

Minute (m)	Number of People (p)
1	38
2	76
3	114
4	152

◀ **Hint** The table shows how many people watched *in all*, not how many people watched during each minute.

Function Rule: $p =$ _____ $\bigcirc$ _____

Ask Yourself

What operation will I use to show the increase in the number of people?

Answer _____

Apply Could you be sure of the pattern if you had only the first two rows of the table? Explain.

_____

_____

Ask Yourself

Will there be more computers or more classrooms?

④ A school is getting new computers that can allow students to share live video with other schools. Every 4 classrooms will share 2 computers. Every 6 classrooms will share 3 computers. Every 8 classrooms will share 4 computers. Every 10 classrooms will share 5 computers. If the pattern continues, how many computers will 32 classrooms share?

Number of Classrooms (x)	Number of Computers (y)

Hint Use the function rule to find the solution.

▶ Function Rule: $y = x \bigcirc$ _____

Answer _____

Relate What number sentence can you write using inverse operations to check that your solution is correct?

_____

_____

⑤ Gia listens to audiobooks. The table lists how much they cost. How many audiobooks can she buy if she has $195.00? How much money will she have left over?

Hint Find a function rule that describes the data.

Number of Audiobooks (a)	Total Cost (c)
2	$42.00
3	$63.00
4	$84.00
5	$105.00

Ask Yourself

How do I find how much money she will have left over?

Function Rule: $c = a \bigcirc$ _____

Answer _____

Conclude How does the function rule explain the pattern?

_____

_____

On Your Own

Solve the problems. Show your work.

⑥ George is setting up tables for a technology conference. The table shows the total number of chairs he uses for a given number of tables. If 114 people are coming to the conference, how many tables will he need?

Tables	Chairs
2	12
3	18
4	24
5	30

Answer _____

Identify What function rule could you use to describe the pattern? How did you find the rule?

_____

_____

⑦ Ms. Day climbs a rock wall at the sports center. Ms. Kwan holds the other end of the climbing rope. Ms. Kwan starts with 4 feet of rope in her hand. The table shows how much rope Ms. Kwan has as Ms. Day climbs. Ms. Day climbs at a constant rate. At what time will Ms. Kwan have 30 feet of rope?

Time (minutes)	Rope (feet)
0	4
7	11
14	18
21	25

Answer _____

Consider Why does finding a pattern help you describe a function rule?

_____

_____

Create   Look back at the problems in this lesson. Write and solve a similar problem that can be solved by finding a pattern.

Strategy Focus
Draw a Diagram

MATH FOCUS: Add and Subtract Integers

Learn About It

▢ Read the Problem ....

> An ancient city was discovered underwater off the coast of Greece. The city is believed to be more than 5,000 years old. Dr. Black and Dr. Costa explore the city. Dr. Black dives to a depth 7 feet below sea level. Then she dives 5 feet deeper. Dr. Costa stops 6 feet above Dr. Black. To what depth does Dr. Costa dive?

Reread Ask yourself questions as you reread again.

• What is the problem about?

_____

• What information is given?

_____

• What do I need to find?

_____

Mark the Text

▢ Search for Information ....

Read the problem again. Circle important words and numbers.

Record Write what you know about the dive.

First, Dr. Black dives _____ feet below sea level.

Then she dives _____ feet deeper.

Dr. Costa stops _____ feet above Dr. Black.

Think about how you can use this data to solve the problem.

Decide What to Do

You know how deep Dr. Black dives. You know how Dr. Costa's position relates to Dr. Black.

Ask How can I find how deep Dr. Costa dives?

- I can use the strategy *Draw a Diagram*.

- I can draw a diagram to show how deep Dr. Black dives. Then I can use the diagram to find how deep Dr. Costa dives.

Use Your Ideas

Step 1 Draw a diagram showing the depth underwater. Use arrows to show how deep Dr. Black dives. First, she dives down 7 feet.

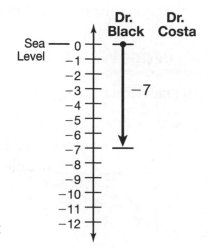

Step 2 Draw an arrow to show that Dr. Black dives down another 5 feet.

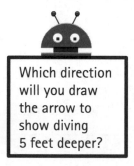

Which direction will you draw the arrow to show diving 5 feet deeper?

Step 3 Draw an arrow showing 6 feet above Dr. Black. This shows how deep Dr. Costa dives.

So Dr. Costa dives to a depth of _____ feet below sea level.

Review Your Work

Read the question again and check that your diagram is correct.

Describe Why is a diagram helpful in solving this problem?

_____

_____

Try It

Solve the problem.

(1) Ruby rides the subway. There are 11 other people in the subway car when she gets on. At the next stop, $\frac{2}{3}$ of the people get off and 7 get on. At the third stop, 3 people get off. What integer expresses the change in the number of passengers between the time Ruby gets on, to just after the third stop?

Mark the Text

Read the Problem and Search for Information

Underline the question you need to answer.

Decide What to Do and Use Your Ideas

You can draw a diagram to keep track of the number of people.

Step 1 Draw a number line. Start at the number of people in the subway car once Ruby gets on. Draw arrows to show that _____ people get off and _____ people get on.

Ask Yourself

What is $\frac{2}{3}$ of 12?

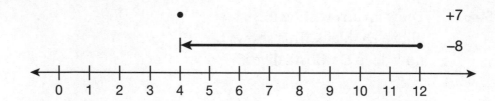

Step 2 Draw an arrow to show that _____ people get off at the third stop. There are _____ people left in the subway car just after the third stop. This is _____ fewer people than just after Ruby got on.

So the integer that expresses the change in the number of passengers is _____ .

Review Your Work

Check that your diagram represents the numbers in the problem.

Identify What is another question you can ask about this problem?

_____

_____

Apply Your Skills

Solve the problems.

② A hotel in Sweden is made entirely of ice and snow. The table shows the temperatures outside the hotel for 5 days. On Monday, the temperature inside the hotel was 8 degrees warmer than it was outdoors. On Friday, the temperature inside the hotel was 5 degrees warmer than it was outdoors. Was it colder inside the hotel on Monday or Friday?

Outdoor Temperature	
Monday	−13°C
Tuesday	−21°C
Wednesday	−37°C
Thursday	−16°C
Friday	−11°C

Friday Outside •

Monday Outside •

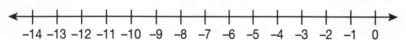

Hint Use arrows to show the inside temperatures on Monday and Friday.

Ask Yourself

If the temperature is warmer, do I move left or right on the number line?

Answer _____

Apply How can you use the diagram to compare the inside temperatures on Friday and Monday?

_____

③ An office worker works in the Rogers Tower. He gets on the elevator at the tenth floor. He rides down 13 floors to Level C. Then he rides up 2 floors. He decides to use the stairs to get to the ground floor. How many floors does he need to walk up or down to get to the ground floor?

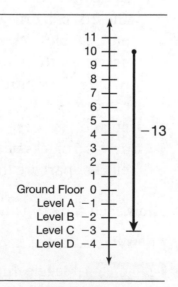

Hint Use the number line to show the directions and how far the elevator moves.

Ask Yourself

Why is zero used to represent the ground floor?

Answer _____

Summarize Explain how you used the number line to solve.

_____

_____

(4) Mrs. Cho has an electronic pass to pay tolls. The amounts of the tolls are shown in the table. One morning, she has $6 left on her pass. She adds money to the pass on her way to work. Then she goes over the Adams Bridge. On the way home, she goes through the Taft Tunnel. When she gets home, she has $8 left on her pass. How much money did she add to her pass?

Bridge or Tunnel	Toll
Adams Bridge	$8.00
Taft Tunnel	$5.00

Ask Yourself

What information is unknown?

Hint Start with the amount that is on the pass in the morning. Subtract the amounts of the tolls first. Then decide how much to add to get $8.

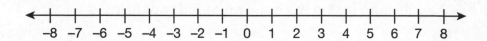

Answer _____

Plan Suppose you know how much Mrs. Cho started with, what she added, and what she had left. You know the toll for the Adams Bridge, but not the Taft Tunnel. How would your diagram change?

_____

_____

(5) Two cabs leave a hotel at the same time with passengers. The first cab drives 2 blocks north. It picks up another passenger there and drives twice as far south. The second cab drives 4 blocks south. It picks up another passenger there and drives 5 blocks north. How many blocks apart are the cabs?

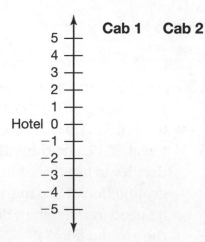

Show each cab's trip on the number line.

Hint Go up on the number line for north, and down for south.

Ask Yourself

What number shows the location of the hotel?

Answer _____

Determine Maggie thinks the cabs are 9 blocks apart. Explain what mistake she might have made.

_____

_____

On Your Own

Solve the problems. Show your work.

6 At 8 A.M., the temperature was −5°F. By 10 A.M., the temperature had dropped 2°. By 1 P.M., the temperature had risen 9° from what it was at 10 A.M. What was the temperature at 1 P.M.?

Answer _____

Relate How could you write an equation to help solve this problem?

_____

_____

7 Randy works at the loading docks. He uses a crane to lift crates from ships to the dock. Today, there are 65 crates to be unloaded. He begins with the crane at ground level. He lowers the crane 14 feet and picks up a crate. He lifts the crate 25 feet and places it on the loading dock. How far does Randy have to raise or lower the crane to get it back to ground level?

Answer _____

Analyze What information is given that is *not* needed to solve the problem?

_____

_____

Create

Look back at the problems in this lesson. Write a similar problem that can be solved by drawing a diagram. Then solve your problem.

Strategy Focus
Make a Graph

MATH FOCUS: Coordinate Graphs and Line Graphs

Learn About It

▢ Read the Problem

Dr. Sanchez is an archeologist. She is making a map of an ancient city found in the middle of a rainforest. The coordinates show where each building appears on the map. Which building is farthest north of the palace?

Building	Location Coordinates
City Gate	(6, 0)
Guard House	(3, 5)
Main Square	(6, 3)
Palace	(2, 3)
Royal Tomb	(7, 9)

Reread Use your own words to tell about the problem.

- What is Dr. Sanchez doing?

_____

- What information is given about the buildings?

_____

- What do I need to find?

_____

Mark
the Text

▢ Search for Information

Read the problem again. Look for important information in the problem and the table.

Record What coordinates are given in the problem?

City Gate: _____

Guard House: _____

Main Square: _____

Palace: _____

Royal Tomb: _____

Think about how you can use this information to solve the problem.

Decide What to Do

You know the coordinates of the locations of the buildings.

Ask How can I find the building farthest north of the palace? Closest to the north of the palace?

- I can use the strategy *Make a Graph*. I can plot the points on a coordinate grid.

- Then I can compare the distances between the palace and the other buildings.

Use Your Ideas

Step 1 Plot the points. Label each building.

> Remember that the *x*-coordinate tells you how far right or left. The *y*-coordinate tells you how far up or down.

Step 2 Look for the building that is farthest north of the palace.

Which buildings are north of the palace?

_____ and _____

Which is farther north? _____

So the _____ is farthest north of the palace.

Review Your Work

Check that you correctly plotted the points on the graph.

Explain How does making a graph help you solve the problem?

_____

_____

Try It

Solve the problem.

Ask Yourself

What does *grow at the same rate* mean?

Mark the Text

① The table shows the height of a giant bamboo plant at different times. If the plant continues to grow at the same rate, how tall will the bamboo be on Day 12? Day 15?

Day	Height (feet)
3	3
6	7
9	11

Read the Problem and Search for Information ....

Identify the information you can use to solve the problem.

Decide What to Do and Use Your Ideas ....

You know the heights of the plant on different days. You can plot the data to make a line graph.

Step 1 Label the *x*-axis with the number of days. Use the *y*-axis to show the height of the bamboo.

Step 2 Plot the data in the table. Connect the points.

Step 3 Extend the line. Mark the point for Day 12 and Day 15.

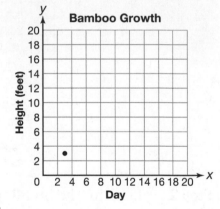

The bamboo plant will be _____ feet tall on Day 12.
It will be _____ feet tall on the Day 15.

Review Your Work ....

Check that you correctly plotted each point on the graph.

Describe How could you use a pattern to solve this problem?

_____

_____

Apply Your Skills

Solve the problems.

2 Dr. Ramirez makes a map of the layers of the rainforest. She plots points where she has seen the animals shown in the table. Which animals did she see in the canopy layer?

Animal	Location
Giant Anteater	(5, 1)
Howler Monkey	(8, 11)
Jaguar	(8, 5)
Morfo Butterfly	(1, 14)
Red-eyed Tree Frog	(2, 4)
Toucan	(4, 12)

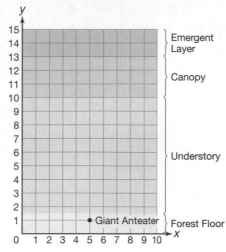

Ask Yourself

To plot a point, do I first move to the right, or up?

Hint Label the location of each animal as you plot it on the graph.

Answer _____

Conclude One student thinks that the howler monkey is in the understory. What mistake might the student have made?

_____

_____

3 Sara takes a walking tour. There are 8 bridges on the tour, all the same distance apart. Sara notes how long the tour has lasted after crossing the 2nd, 3rd, and 6th bridges. At this rate, when will Sara pass the 8th bridge?

Number of Bridges	Time (minutes)
2	30
3	45
6	90

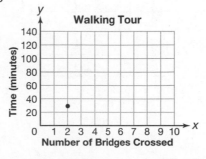

Ask Yourself

What is the y-coordinate of the point whose x-coordinate is 8?

Hint You can give your answer in hours or minutes.

Answer _____

Identify What is another question you could ask?

_____

_____

Ask Yourself

How will plotting the points help me find which location is farthest away?

Hint Remember to show the hotel on the map. ▶

4 Bobby is on a trip to the rainforest for 10 days. He makes a map showing places in the table he wants to visit. Bobby's hotel is at (4, 3). Which location is farthest from his hotel?

Site	Location
Butterfly Garden	(6, 2)
Waterfall	(9, 7)
Horseback Riding	(1, 5)
Bike Rentals	(3, 5)

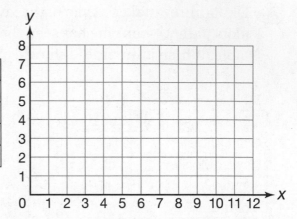

Answer _____

Determine What information is given that is *not* needed to solve?

_____

_____

Hint First you need to find the number of species Dr. Witt studied in Week 7 and in Week 4. ▶

5 Dr. Witt records the total number of plant species she has studied. The pattern in the table continues. About how many more species has Dr. Witt studied at the end of Week 7 than at the end of Week 4?

Ask Yourself

How should I label the x-axis?

Week	Total Number of Species
1	10
2	17
3	24
5	38

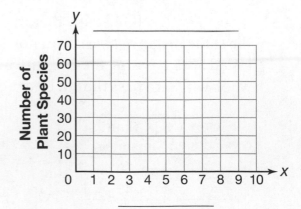

Answer _____

Examine What words tell you that you do *not* need an exact answer?

_____

_____

On Your Own

Solve the problems. Show your work.

(6) The map shows the average rainfall in a rainforest during January. There are three sites to be plotted on the map. Site A is at (2, 9). Site B is at (10, 7). Site C is at (5, 2). Which site had the most rain? The least rain?

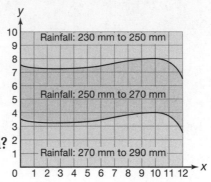

Answer _____

Analyze What mistake might you make if you only look at the position of the points and not the labels on the graph?

_____

_____

(7) Neil collects rainforest trading cards. The number of cards he has at the end of several months is shown in the table. The number of cards he has increases at the same rate each month. How many cards will Neil have at the end of the sixth month?

Month	Number of Cards
1	22
3	34
4	40

Answer _____

Plan How could you use a graph of the points in the table to find the number of cards Neil will have at the end of one year?

_____

_____

Look back at the problems in this lesson. Write a similar problem that can be solved by making a graph. Then solve your problem.

Create

In this unit, you worked with four problem-solving strategies. You can often use more than one strategy to solve a problem. So if a strategy does not seem to be working, try a different one.

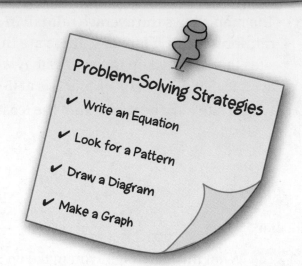

Problem-Solving Strategies
- ✔ Write an Equation
- ✔ Look for a Pattern
- ✔ Draw a Diagram
- ✔ Make a Graph

Solve each problem. Show your work. Record the strategy you use.

1. A robot moves along a straight path. The robot moves forward 5 spaces. Then it moves back twice as far. The robot moves back 2 more spaces. Then it moves forward 3 spaces. What integer represents the robot's location on the path compared to where it starts?

2. Marc and Teresa go bird watching. Marc sees 24 birds. He sees 15 fewer birds than Teresa. How many birds do they see in all?

Answer _____

Strategy _____

Answer _____

Strategy _____

3. Fred plots these points on a coordinate grid and connects them in order.

$$(1, 4), (2, 1), (6, 1), (5, 4)$$

What shape does Fred make?

Answer _____

Strategy _____

4. Jay is saving money to buy a new MP3 player. It costs $45. The table shows how much Jay has saved each week. If the pattern continues, at the end of which week will Jay have enough money to buy the MP3 player?

Week	Total Amount Saved
1	$6
2	$12
3	$18

Answer _____

Strategy _____

5. On Saturday, 42 inches of snow fell. That is 3 times the amount of snow that was already on the ground. How much snow was already on the ground?

Answer _____

Strategy _____

Explain how you can use a different equation to check your answer.

Solve each problem. Show your work. Record the strategy you use.

6. Ryan records the weight of his puppy at different ages. If his puppy continues to grow at the same rate, how much will it weigh at 10 weeks?

Age (weeks)	Weight (pounds)
2	3
5	9
7	13

Answer _____

Strategy _____

7. The temperature in the morning is −7°F. The temperature rises 10°F. Then it falls 5°F. What is the temperature now?

Answer _____

Strategy _____

8. The nature club is having a recycling drive. At the start of the drive, they already have 4 cans. The diagram shows how many cans they collect as time passes. If the pattern continues, how many cans will they have in 30 minutes?

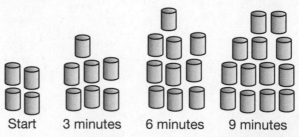

Start   3 minutes   6 minutes   9 minutes

Answer _____

Strategy _____

Explain how you could use the pattern to find the number of cans collected in any given number of minutes.

9. Gerry makes a map of his town. His apartment is located at (4, 5). The school is at (2, 1). The library is at (4, 1) and the town hall is at (1, 5). Which building is closest to Gerry's apartment?

Answer _____

Strategy _____

10. A seal is swimming along the surface of the water. It dives to a depth of −7 meters. It dives down another 15 meters. Then it rises half the total distance to the surface. At what depth is the seal?

Answer _____

Strategy _____

Write About It

Look back at Problem 4. Justify your answer.

_____

_____

_____

Work Together: Make a Graph

Three springs are hanging from a metal beam. Weights are added to the bottom of each spring. The chart shows the length of each spring before and after the weights are added.

Plan
1. Choose one of the springs. Make a graph to show how the length changes as mass is added.

2. Look at the patterns in the change of the lengths of each of the springs. Predict the length of each spring when a 5-kilogram weight is added.

Length of Spring (cm)			
Mass (kg)	Spring 1	Spring 2	Spring 3
0	3	3	5
1	5	6	7
2	7	9	9
3	9	12	11

Decide Suppose you have another spring. Choose a length and a pattern in the way it changes as weights are added.

Create Make a graph showing the relationship between the length of your spring and the amount of mass added.

Present As a group, share your graph with the class. Discuss the pattern in the way the length changes as weights are added and tell what function rule describes the relationship.

Math Vocabulary Activities

On the next six pages are some of the math terms you have worked with in each unit.

You can cut these pages to make vocabulary cards. The games and activities below can help you learn and remember the meaning of these important terms.

Try This!

▶ Complete the activity on the back of each vocabulary card. Use a separate sheet of paper. Discuss your work with a partner or in a small group.

▶ Work with a partner. Take turns. One person chooses a vocabulary card and shows the front of the card. The other person gives the definition. Check to see if your partner was correct by looking at the back of the card.

▶ Draw a picture or write an example of each term on a separate card. Then have a partner match your example or picture to a vocabulary card.

▶ Play a matching game with a partner or a small group.

Make your own sets of cards. Make one set for each term. Write a term on one side and leave the other side blank. Lay out the cards in this set in rows, facedown. Make another set for each definition. Write a definition on one side and leave the other side blank. Lay out the cards in this second set in rows, facedown and separate from the first set.

Take turns. The first player turns over two cards, one from each set. If the cards show a word and its matching definition, the player keeps them and takes another turn. If the word and definition do not match, place the cards facedown where they were, and it is the next player's turn. The player with the most matched pairs wins the game.

Math Vocabulary

difference

15 − 6 = 9
↑
difference

divisor

60 ÷ 2 = 30
↑
divisor

factor

15 × 2 = 30
↑   ↑
factors

multiple

Multiples of 4:
0, 4, 8,
12, 16 …

pattern

2, 4, 6, 8, 10, …

product

15 × 2 = 30
↑
product

quotient

60 ÷ 2 = 30
↑
quotient

remainder

10 ÷ 6 = 1 R4
↑
remainder

decimal point

0.25
↑
decimal point

denominator

$\dfrac{3}{4}$
denominator

like denominators

$\dfrac{1}{4}$ $\dfrac{3}{4}$
like denominators

mixed number

fraction
↓
$1\dfrac{1}{2}$
↑
whole number

numerator

numerator
$\dfrac{3}{4}$

simplest form

$\dfrac{8}{12} = \dfrac{2}{3}$
↑
simplest form

unlike denominators

$\dfrac{1}{4}$ $\dfrac{3}{7}$
unlike denominators

whole number

0, 1, 2, 3 …

Math Vocabulary

multiple

A multiple of a number is the product of that number and some whole number.

Write a sentence using the word.

factor

a number multiplied by another number to get a product

Write an example of the word.

divisor

the number by which another number is divided

Write an example of the word.

difference

the result of subtracting one number from another number

Write an example of the word.

remainder

a whole number that is left over after one whole number is divided by another whole number

Give an example of the word.

quotient

the result of one number being divided by another

Write an example of the word.

product

the result of two or more numbers being multiplied together

Write an example of the word.

pattern

objects or numbers arranged according to a rule or rules

Write an example of a number pattern.

mixed number

a number containing a whole number and a fraction

Write an example of the word.

like denominators

denominators that are exactly the same in two or more fractions

Write two fractions that have like denominators.

denominator

the number in a fraction below the bar that tells how many equal parts there are in a whole or a set

Give an example of the word.

decimal point

a dot separating the ones and tenths places in a decimal number

Write two different examples of the word.

whole number

one of the numbers 0, 1, 2, 3, and so on

Give an example of a whole number and an example of a number that is not a whole number.

unlike denominators

denominators in two or more fractions that are different

Write two fractions that have unlike denominators.

simplest form

A fraction is in simplest form when the greatest common factor of the numerator and denominator is 1.

Write an example of the word.

numerator

the number in a fraction above the bar that tells the number of equal parts of the whole or set that you are talking about

Write two different examples of the word.

Math Vocabulary

decimal

10.75
⎵
decimal

discount

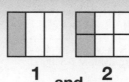
BOOK SALE
$20 Regular Price
10% off
→ discount

equivalent fractions

$\frac{1}{3}$ and $\frac{2}{6}$
are equivalent fractions.

equivalent ratios

stars : circles
☆☆○○○○

2:4 and 1:2
are equivalent ratios.

fraction

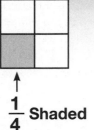

$\frac{1}{4}$ Shaded

percent

75%

rate

Rate:

$\frac{\text{distance}}{\text{time}}$ → $\frac{\text{60 miles}}{\text{1 hour}}$

ratio

2 to 4

2:4

$\frac{2}{4}$

area

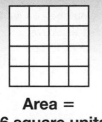

Area =
16 square units

circumference

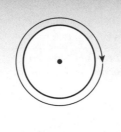

diameter

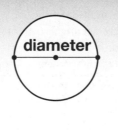
diameter

intersect

These two lines intersect here.

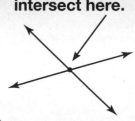

parallel

perimeter

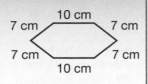
10 cm
7 cm   7 cm
7 cm   7 cm
10 cm

Perimeter =
48 centimeters

perpendicular

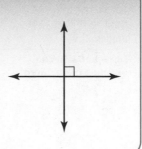

radius

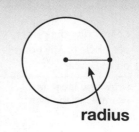
radius

205

Math Vocabulary

Unit 3

equivalent ratios

ratios that are represented by equivalent fractions

Give two examples of equivalent ratios.

equivalent fractions

fractions that name the same amount

Give two examples of equivalent fractions.

discount

an amount or a percent to be subtracted from a regular price

Write two sentences using the word.

decimal

a number with one or more digits to the right of a decimal point

Write two different examples of the word.

ratio

a comparison of two numbers using division

Write an example of the word.

rate

a ratio that compares measurements or amounts

Write an example of the word.

percent

a ratio that compares a number to 100 using the symbol %

Write a sentence using the word.

fraction

a number that names a part of a whole or part of a group

Write and label an example of the word.

Unit 4

intersect

to meet or cross at a point

Draw two different examples of the word.

diameter

a line segment connecting two points on a circle that passes through its center

Draw and label an example of the word.

circumference

the distance around a circle

Draw and label an example of the word.

area

the number of square units that cover a surface without overlap

Draw an example of the word.

radius

a line segment connecting the center of a circle to any point on the circle

Draw and label an example of the word.

perpendicular

intersecting at right angles

Draw and label an example of the word.

perimeter

the distance around a figure

Draw and label two different examples of the word.

parallel

always the same distance apart

Draw an example of the word.

Math Vocabulary

compound event

Tossing heads and spinning a 6 is a compound event.

dependent events

Choose a white marble.
Don't replace it.
Then choose another white marble.

independent events

Choose a white marble.
Replace it.
Then choose another white marble.

mean

$$\frac{1 + 2 + 4 + 5}{4} = 3$$

↑ mean

median

25
31
78 ← median
84
99

mode

15
13
20
28
7
13

mode is 13

outcome

3 is the outcome of the spin.

range

1, 2, 4, 4, 8, 9

9 − 1 = **8**

↑ range

coordinate grid

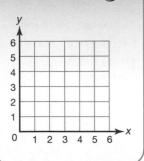

equation

$x + 8 = 18$

function

output input
↓   ↓
$y = x + 3$

This is an example of a function.

function table

Input	Output
x	y
0	3
1	4
2	5
3	6
4	7
5	8

integer

… −2,
  −1,
   0,
   1,
   2, …

inverse operation

$2 \times 5 = 10$
$10 \div 5 = 2$

x-axis

x-axis

y-axis

y-axis

Math Vocabulary

mean

the sum of a set of numbers divided by the number of numbers

Use numbers to show the word.

independent events

Two events are independent events if the occurrence of one of the events does not affect the probability of the occurrence of the other event.

Give an example of independent events.

dependent events

Two events are dependent events if the occurrence of one of the events affects the probability of the occurrence of the other event.

Give an example of dependent events.

compound event

a combination of two or more events

Give an example of a compound event.

range

the difference between the least value and the greatest value in a set of numbers

Write a list of numbers and tell the range.

outcome

a possible result of an experiment

Write a sentence using the word.

mode

the number or numbers that appears most often in a set of numbers

Use numbers to show the word.

median

the middle number in a set of numbers when the numbers are put in order

Use numbers to show the word.

function table

a table that matches each input value with one output value

Draw an example of the word.

function

a rule in which the input number can result in only one output number

Write an example of the word.

equation

a statement that two quantities are equal

Write two different examples of the word.

coordinate grid

a grid formed by an x-axis and a y-axis

Write a sentence using the word.

y-axis

the vertical axis on a coordinate grid

Write a sentence using the word.

x-axis

the horizontal axis on a coordinate grid

Write a sentence using the word.

inverse operation

an operation that undoes another operation

Write two examples of inverse operations.

integer

a number that is a whole number or the opposite of a whole number

Write two different examples of a whole number and its opposite.